Carpentry

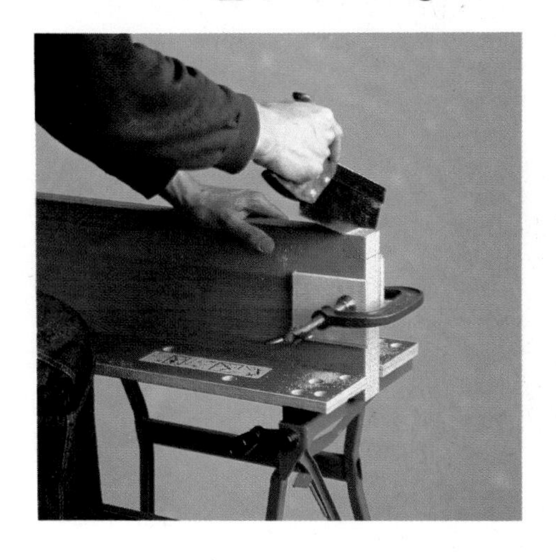

Carpentry

Alec Limon and Paul Curtis

Consultant Editor Mike Trier

OCTOPUS BOOKS

Contents

This edition published 1987
by Octopus Books Limited
59 Grosvenor Street, London W1

ISBN 0 7064 2868 4

Printed in Hong Kong

Introduction

Much of the satisfaction of carpentry is measured by the quality of the end result, and this book sets about helping you to achieve a good finished product and ensure that satisfaction.

An important starting point is to buy good quality materials and make yourself aware of possible faults as, without this knowledge, the quality of a carefully constructed project can be a disappointment. Standard timber sizes vary slightly from one supplier to another, so it is as well to check those of the timber you are using before working out critical dimensions. Also bear in mind that hardwoods have greater strength than softwoods of the same section and you can therefore use smaller hardwood sections to give a less bulky appearance to your constructions.

The use of elaborate and expensive tools alone will not ensure good workmanship. The secret of success is knowing which is the right tool to use for a particular job, and how to use it. In this book you will find details of the basic tools and techniques for using them, as well as some of the more specialised tools, including some which can be home-made.

Having selected suitable materials and the necessary tools, the next important step is setting out the parts carefully and accurately, using the correct methods. Always remember to check and double check all dimensions before proceeding – a pencil line is easy to erase and reposition, but undersized pieces of timber are costly and cause frustration. A good carpenter will only use a cutting list as a guide: at each stage of the construction, he or she will measure the dimensions of the parts to fit exactly, and cut so as to leave the guideline standing.

All this information will be helpful if you are working from a set of plans and instructions. However, if you wish to design your own projects, you will need to know about different principles and methods of construction. Conventional woodworking joints and fixings, techniques for working with man-made boards and box construction are also described.

All homes need a variety of storage units designed to suit different situations, such as under the stairs or along a living-room wall. A variety of designs are covered here, including exploded drawings of some typical constructions, a selection of fittings available, and techniques for hanging a door. Do remember when fixing a unit to a wall to check first for concealed pipes and wiring. Small electronic detectors are available which make this task easy, but a general rule is never to make fixings vertically above or below electrical fittings, and to avoid the vicinity of pipe runs near plumbing appliances.

A superb finish can be the highlight of an otherwise everyday article. Different finishes are described for showing off the grain of the timber to the full or for colouring it to suit your taste. And, to enable you to put into practice some of the techniques described, you will find some excellent furniture projects with working drawings and instructional text.

Remember that SAFETY WARNINGS and MANUFACTURERS' INSTRUCTIONS are designed for your safety and health. Use the proper equipment and wear the correct clothing – never wear a tie. Do take particular care with any electrical work, it is never trivial, and when working with chemicals. Always protect your eyes with goggles when the situation demands it. Never interfere with, or remove, safety guards on power equipment. Concentrate on safety at all times and do not rush work – it is not worth it.

6 BUYING TIMBER

Ignorance of the nature of the materials you are using can cause a project to fail. Timber is a natural material and it is necessary to know something about the way it grows in order to understand how it can best be used and how it will behave in use.

Briefly, trees grow by adding new layers round the outside of the trunk, just underneath the bark. This makes the well-known annual rings. The sap which feeds these new layers also passes through to the centre or heartwood along the medullary rays which we call the figuring when the wood has been cut into planks. This means that the first few centimetres of the tree immediately under the bark are not fully grown cells and the mature timber is found towards the middle of the tree. The actual centre is known as the pith. These distinct areas behave differently in use.

Sapwood, that is the outer, immature wood, shrinks more than the inner heartwood. It is therefore more liable to twist. As shrinkage takes place round the direction of the annual rings from the outer part of the tree towards the central heart, if the tree is simply cut straight through into wide planks they will contain more sapwood on one side than the other, and will warp away from the heart side as the sapwood dries out and shrinks more than the heartwood. The pith should be avoided for all but the least important jobs because it is the most likely to twist as it dries or seasons.

When buying timber, as well as looking for the obvious defects such as large knots, shakes and cracks, and at general appearance, look also at the end grain, as it will tell you a lot about the future movement of the wood. The best pieces have the annual-ring marks fairly vertical across the narrowest width. Square timbers should not have the rings running diagonally as this will distort the shape.

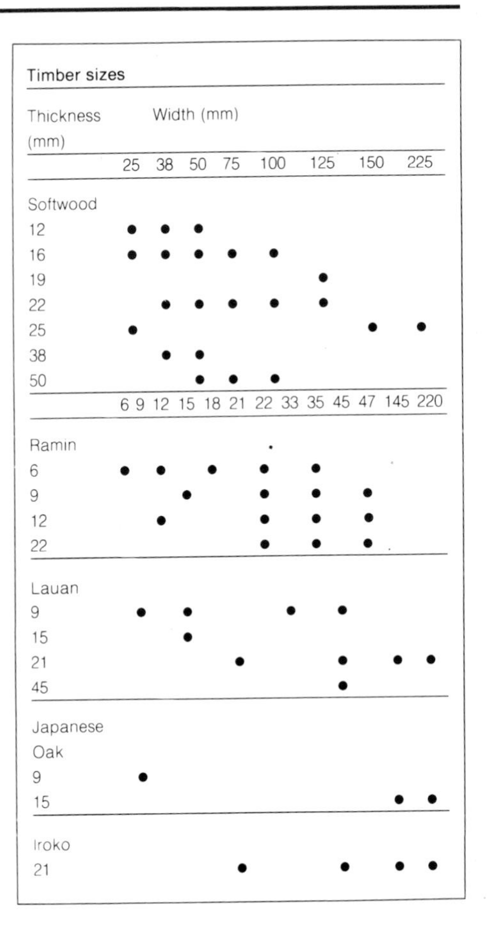

Timber sizes

Thickness (mm)	Width (mm) 25	38	50	75	100	125	150	225
Softwood								
12	•	•	•					
16	•	•	•	•	•			
19						•		
22		•	•	•	•	•		
25	•						•	•
38		•	•					
50			•	•	•			

Thickness (mm)	6	9	12	15	18	21	22	33	35	45	47	145	220
Ramin													
6	•		•		•		•		•				
9				•			•		•		•		
12			•				•		•		•		
22							•		•		•		
Lauan													
9		•		•				•		•			
15				•									
21						•				•		•	•
45										•			
Japanese Oak													
9		•											
15												•	•
Iroko													
21						•				•		•	•

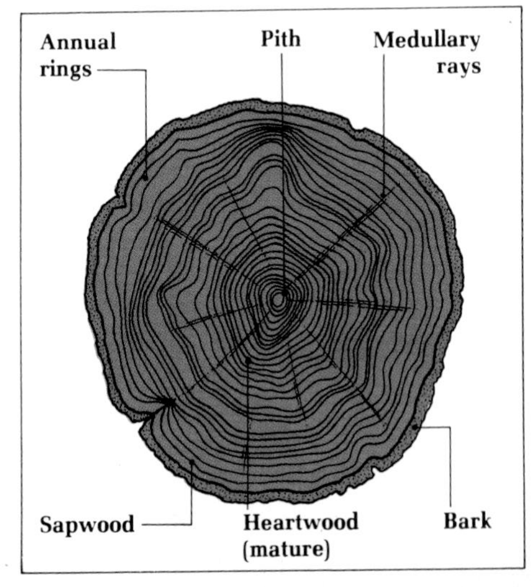

Above right *Table of commonly available sections of softwoods and hardwoods.*

Right *Cross-section through a tree showing important features.*

Points to watch when choosing timber

- Choose timber which is both straight in its length and free from twist.
- Avoid large knots or dry dead knots which are likely to fall out of the board. Small well-spaced knots are acceptable.
- Choose vertical end grain.
- The closer the growth rings, the denser and more durable the timber.
- As well as obvious splits, or shakes as they are called, look for cup shakes which follow the curve of the growth rings and are equally weakening.
- Avoid waney edge; this is the outer part of the tree and may still have the bark attached. It would need preservative.
- Sapwood of some of the softwoods is prone to blue staining. This is a type of mould and not serious, but because it indicates immature wood, a preservative should be applied for protection.

Other diseases to look for (which mainly apply to secondhand timber) are:

- Wet rot, which will die when the water is removed.
- Dry rot, in which the wood is dry, brown, crumbly and cracked into squares. This, like woodworm infected timber, should never be brought into the house.
- Woodworm infected timber, recognised by the flight holes of the beetle which are about 3mm in diameter.

a *Wider dark rings, higher strength.*
b *Waney edge with bark removed.*
c *Drying shakes reduce strength.*
d *Rupture from sharply sloping grain.*

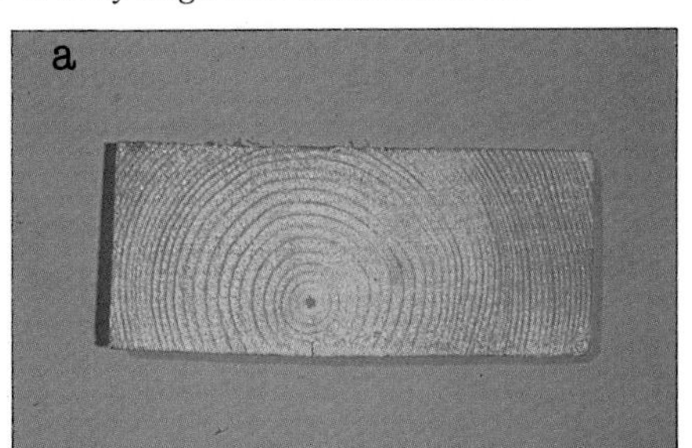

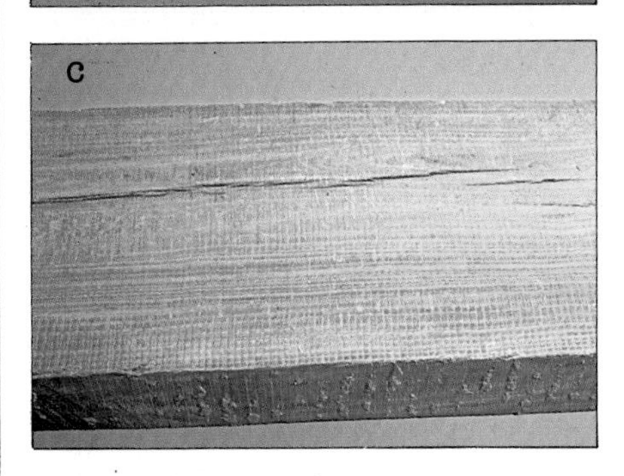

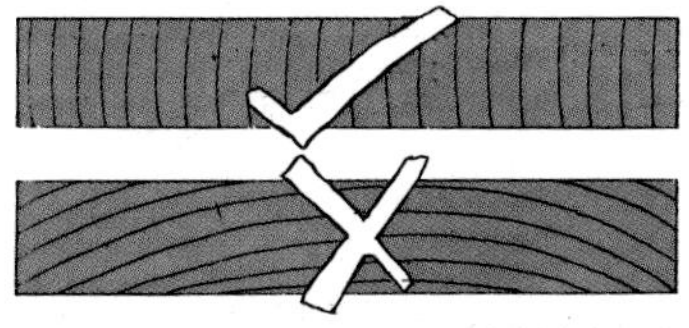

Left *Good and bad end-grain patterns for rectangular and square sections of timber.*

Usage of timber

Timber not only dries and shrinks but also takes up moisture and swells. Its tendency to change shape affects the choice of wood and the way it is used. For example, you should always put the timber heart side up for table tops and the tops of wooden stools, and table tops should not be fixed firmly, only buttoned or attached with slotted plates to allow for this movement.

Further, if wide boards are to be made up using a number of narrow boards edge-jointed together, it is best to arrange the boards so that they are alternately heart side up and heart side down. This will cause them to pull against each other and in this way help to hold them flat. The battens underneath, which are used to hold the boards, must be screwed through slots, not through tight screw holes. In this way the whole top can adjust itself to the atmospheric conditions.

These comments apply to both hardwoods and softwoods, although the movement of softwoods is greater than that of hardwoods. However, the movement of the latter can often have a more damaging effect because these timbers are used for furniture show-woods.

Hardwoods and softwoods

The terms hardwood and softwood are based on the structure of the wood and the characteristics of the trees. In general, softwoods are from trees with small, pointed leaves of the cone-bearing and evergreen types, while hardwoods are from trees which shed their leaves in autumn. For practical purposes, softwoods are used for general joinery and structural work while hardwoods are generally used for furniture and show woods.

Buying boards

There are three main types of manufactured wooden boards – hardboard, chipboard and plywood (including blockboard). They are sold by the square metre, hardboard and plain, standard chipboard being cheaper than veneered chipboards, plywoods and blockboards of the same thickness.

Hardboard is shredded timber which has been pulped and remade as large sheets. It can be plain or patterned and there is an oil-tempered type which is for use externally. Hardboard is used for the base of boards finished with a woodgrain pattern or tiled finish. Fretted pattern boards are also made and these make good decorative grilles.

Chipboard is made from timber chipped into flakes and then pressed, with a resin binder, to make a large sheet of material. It is available in stock sizes of length and width, and with a range of timber-veneer or plastic facings. Special grades are available for flooring.

Plywood is made in a very wide range of sizes and thicknesses. The number of plies varies according to the thickness required which can be from about 4mm to 25mm. All kinds of facing veneers are available from plain birch to mahogany and teak. For panelling there is a planked-pattern finish as well as plain wood.

Laminated boards consist of a core of wood strips glued or laminated into one solid piece, faced with veneer. They are obtainable in similar finishes to plywood. Pineboard also comes under the general heading of laminated boards. It is made from strips of pine edge-glued together and is available in a range of widths and lengths.

Usage of boards

Hardboards and softwood-faced plywoods can be used for similar purposes. When buying them take care that the sheets are completely flat, because once they have taken up a curved or distorted shape they are difficult to straighten.

Chipboards and blockboards also have similar uses and for the same reasons care must be taken that they are flat. Both boards are stable, but as shelving, blockboards will hold greater weights over larger spans. Blockboard has a surface which will accept a finish without further work, but chipboard needs to have a decorative surface applied to it.

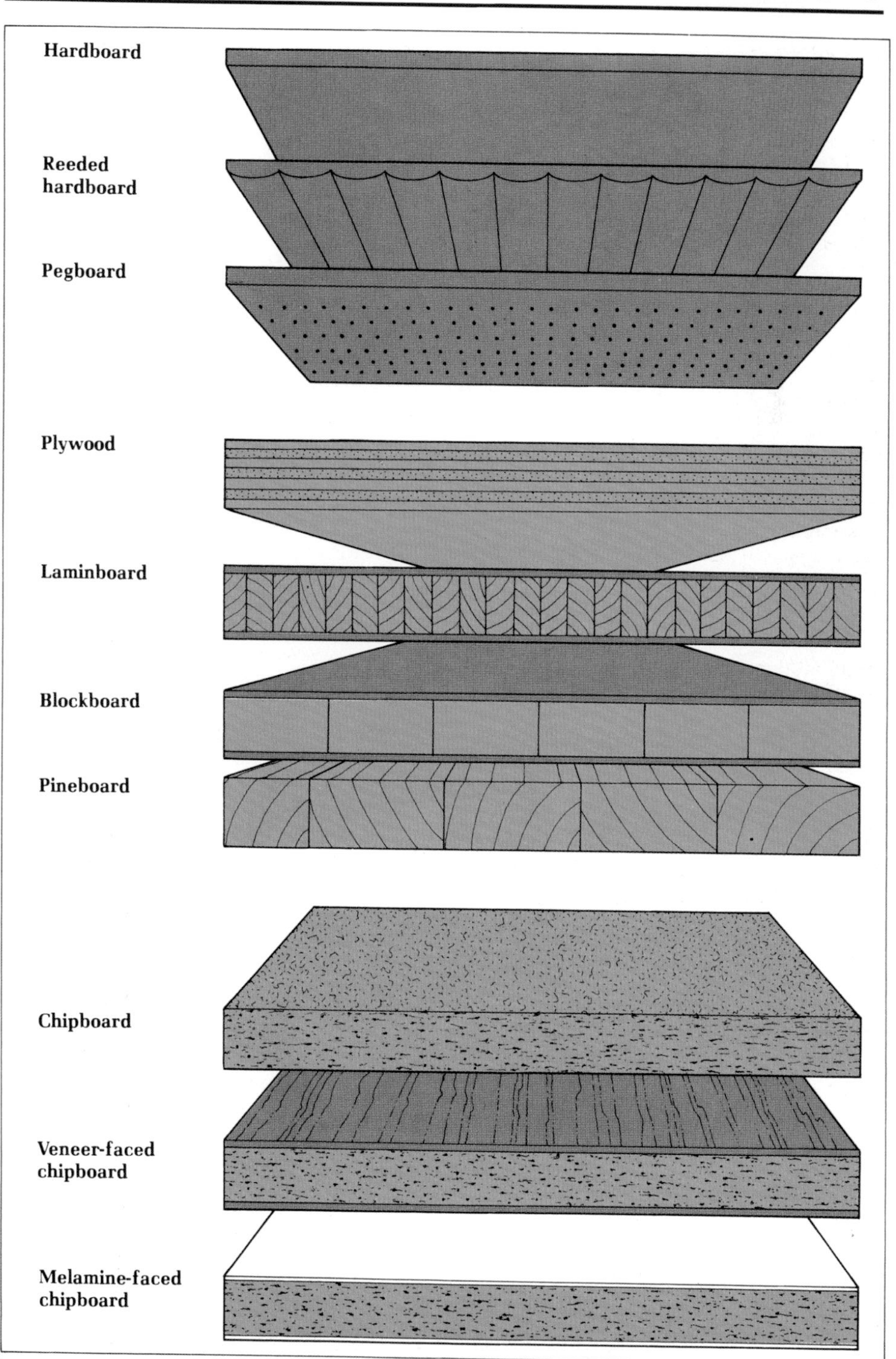
Hardboard
Reeded hardboard
Pegboard
Plywood
Laminboard
Blockboard
Pineboard
Chipboard
Veneer-faced chipboard
Melamine-faced chipboard

Exactly what constitutes a basic tool kit will depend to some extent on what kind of work you intend to do.

Measuring and setting out

In furniture construction and general workshop joinery there is always measuring to be done, whatever the job. For setting out on the bench the best measuring tool is a good quality folding rule. For measuring around the house use a steel tape. A two or three metre long model will meet most requirements. Get a type which has both metric and imperial markings but be sure you only work in one or other system.

A square is needed to mark wood for cutting and although a try-square with a blade 200 or 230mm long is used for benchwork, a 300mm adjustable combination square and mitre square with a sliding blade is more useful for other work. For marking out the cutting lines you will need a setting-out, or marking, knife and a pencil sharpened to a chisel point (this is more accurate than a pointed one).

Cutting

To cut timber you will need a panel saw 500 or 550mm long with 10 teeth per 25mm. This is a suitable size for general work. For bench work you need a tenon saw. This should have 12 or 14 teeth per 25mm. When sawing, the saw should not be forced to cut but pushed evenly so that it can cut at its own speed.

Shaping

Bevel-edged chisels are the best choice for bench work, but for general purposes use a firmer chisel which has square sides and is stronger. Widths of 19 and 6mm would enable you to make a start.

There is a wide range of planes available, from small block-planes up to the long trying-planes used for obtaining straight edges. The latter is not often needed now as most timber can be bought planed straight.

Selection of basic joinery tools:
a *Firmer chisels;* **b** *Bevel-edged chisels;* **c** *Oilstone;* **d** *Steel tape;* **e** *Tenon saw;* **f** *Panel saw;* **g** *Smoothing plane;* **h** *Slotted and Supadriv screwdrivers;* **j** *Bradawl;* **k** *Auger bits;* **l** *Twist bits;* **m** *Sanding block;* **n** *Electric drill;* **p** *Mallet;* **q** *Pincers;* **r** *Claw hammer;* **s** *Warrington hammer;* **t** *Swing brace.*

A block plane comes in very handy for rough shaping; buy one which has screw adjustment for the blade. It can be held in one hand, with the forefinger resting on the front knob of the plane, and the top of the wedge and the cutting blade in the palm of your hand. The larger, 250mm smoothing plane requires two hands to operate it successfully: one holds the handle at the back and the other holds the front knob. The hand on the knob holds the plane level at the start of the cut and raises it slightly at the end of the cut. Failure to lift the plane in this way will cause the end of the timber to be made slightly rounded as the plane drops as it reaches the end of the cut.

With both these types of plane, adjustment of the blade is made by turning a knob under the blade. This is just in front of the rear handle of the smoothing plane. Side adjustment is made by a lever fitted under the blade.

Sharpening

An oilstone is required for sharpening these tools and a medium grade provides a suitable edge for a beginner, but later you may want a finer stone in order to hone a keen edge on tools used for benchwork. The cutting edges of chisels and planes have two bevels, one at about 25 degrees made by the grindstone, and the other at about 35 degrees which is the sharpening bevel. Do not raise the handle of chisels and other blades too high when sharpening them or the cutting bevel will become too steep, reducing the ability to cut cleanly.

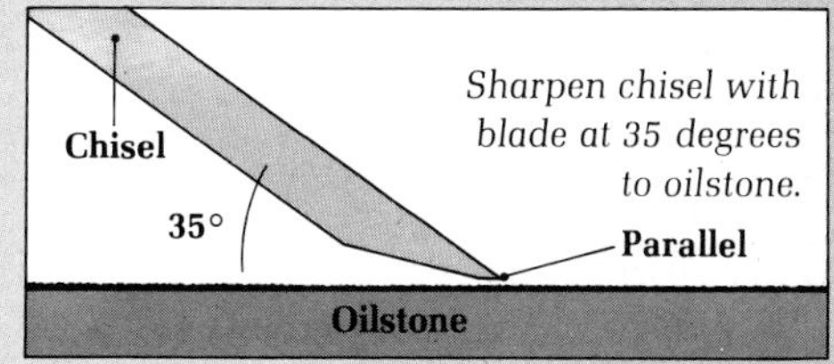

Sharpen chisel with blade at 35 degrees to oilstone.

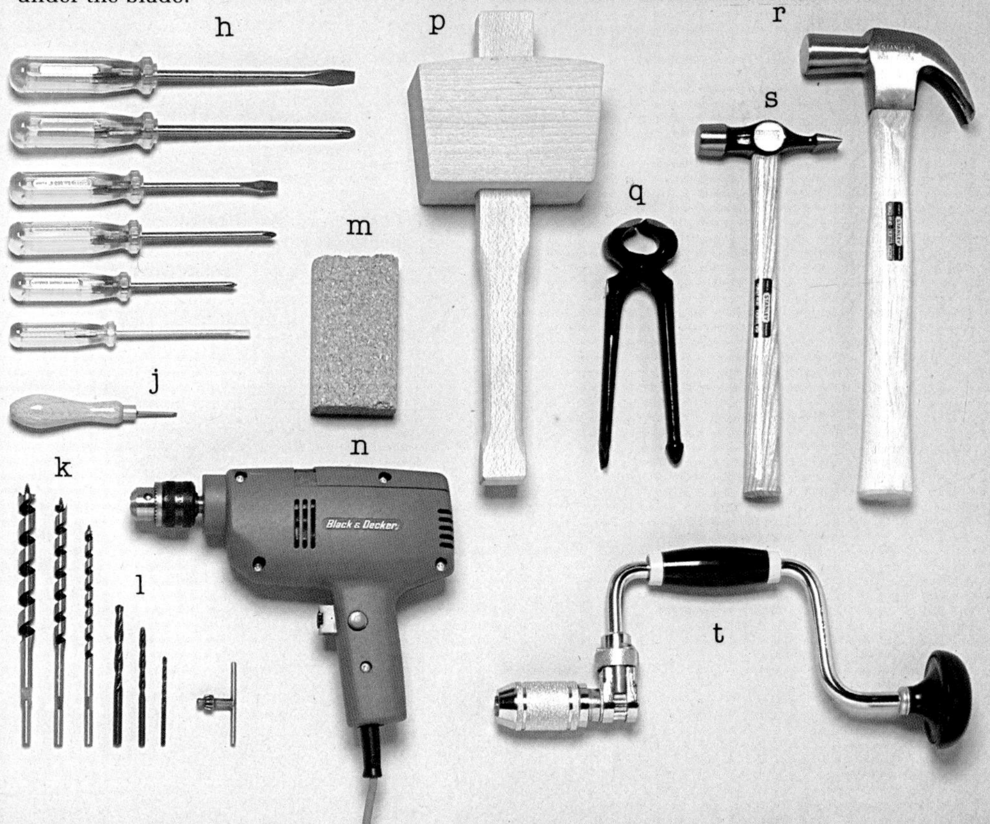

For right-handed people, hold the blade or handle firmly in the right hand, with the left hand resting on the flat blade steadying it and providing an even pressure. This is particularly important with wide plane blades as the edge must be kept square. However, the corners of the plane blades have to be rubbed off a little to prevent them digging into the wood and making marks on the surface.

Use the full width of the stone when sharpening chisels as it will otherwise wear hollow in the middle and be unable to sharpen wide blades properly.

Drilling

Drilling holes, whether for fixing or for assembling projects, is another essential part of woodwork. An electric drill is a boon for holes up to 9mm, although a simple hand-drill (or wheel brace) will do the job. For holes above this size you have a choice of using flat-bits in an electric drill or buying a carpenter's swing brace which will take a wide variety of auger bits for making deep holes of almost any size. Adjustable bits are available, and for work in confined spaces, a ratchet brace is useful.

Assembling

When it comes to assembling furniture and fittings you will need a hammer for driving nails, and a screwdriver or two. Suitable screwdriver lengths would be 150mm and 230mm. You will need tips suitable for cross-head (Supadriv) as well as the traditional slotted-head screws. A bradawl is essential for making a small hole for starting the screws.

A Warrington-pattern hammer is best for bench use and it should be about 12oz in weight. This is the type with the cross pein at the opposite side of the head to the striking face. This chisel-shaped pein is used for starting small panel-pins held between the finger and thumb. The claw hammer, much favoured by carpenters, is ideal for heavier fixings around the house and is of greater weight, about 20oz. If you have only a Warrington-pattern hammer you will need a pair of pincers to pull out nails and tacks.

Assembling furniture made with conventional woodworking joints calls for a mallet. This tool used to be essential for use with chisels, but the modern plastic handle is able to withstand the use of a hammer. The wider head of the mallet and its softer nature makes it less likely to mark the timber when projects are being assembled. The main difference in using these tools is that the hammer is swung with a wrist action whereas the mallet is swung from the elbow.

Finishing

When projects have been completed there remains the finishing. Whether this is to be paint, varnish or polish a certain amount of preparation is necessary. Hardwoods are the most difficult to prepare because the grain may be twisted, figured or reversed and though this provides a pleasant appearance, it does take some effort to produce a good finish. When this kind of grain is to be polished or varnished it has to be scraped first.

An ordinary scraper is simply a rectangle of metal and it is sharpened with a hard steel rod by first drawing it along the edge of the scraper, square to the sides, and then tilting the rod at a slight angle and drawing it along the edge of the scraper again. This has the effect of first creating a burr on the metal then bending the burr over slightly.

The scraper is held upright with the thumbs at the back; it is pushed forward over the surface of the wood where the burr will take off very fine shavings. It takes more than a little practice to sharpen and use this tool.

The alternative is abrasive paper. There are many grades of abrasive papers and many different types of grit. Very fine ones are used for preparing wood for polish and varnish, but the coarser grades of glass-paper are all that are needed to prepare timber for painting. A comfortable sanding block either cork or cork-faced is ideal for holding these papers on flat surfaces.

If you have a workbench you will be able to fix a vice to it so that you can hold timber firmly while working on it. When it comes to assembling the parts you will need other holding devices.

Bench hook

The simplest of holding devices is the bench hook. This can be made out of a piece of 75 × 50mm timber about 225mm long. It is cut out at each side so that a 50 × 25mm block is left at each end, on opposite sides of the wood. In use, it is placed on the bench so that one of the blocks is downward and hooks against the bench top, and the other faces upward so that the timber to be cut can be held firmly against it.

An alternative form of this hook is made from a broad piece of wood approximately 150mm wide, with a block screwed to opposite sides at opposite ends. Make these blocks about 25mm shorter than the width of the board and you can cross-cut timber held against the stops and when the saw comes through the wood it will not damage the bench top.

Whether you use patent cramps, traditional cramps or home-made cramps using folding wedges, the important part of the operation is holding the members of the structure tightly until the glue has set or screws, nails or other permanent fixings have been employed.

Sash cramps

One of the most useful devices is the sash cramp of which there are patent types as well as the simple, traditional cramp. This consists of a long metal bar with a screw-adjustable jaw at one end and a movable jaw which slides along the bar. The latter is fixed at the required point by a pin which passes through holes in the bar. Two of these cramps are needed for pulling up the joints of frames, but four or even more would be ideal for furniture making if you are using traditional cabinet-making methods of construction.

Frame and web cramps

For cramping chairs, frames or other similar constructions there are frame and web cramps. The frame cramp consists of a length of strong nylon cord with four corner blocks and a cleat. The corner blocks are put in position and the cord is passed round the frame over the blocks and tightened on the cleat. This puts an even pressure on all four corners at the same time. The web cramp consists of nylon

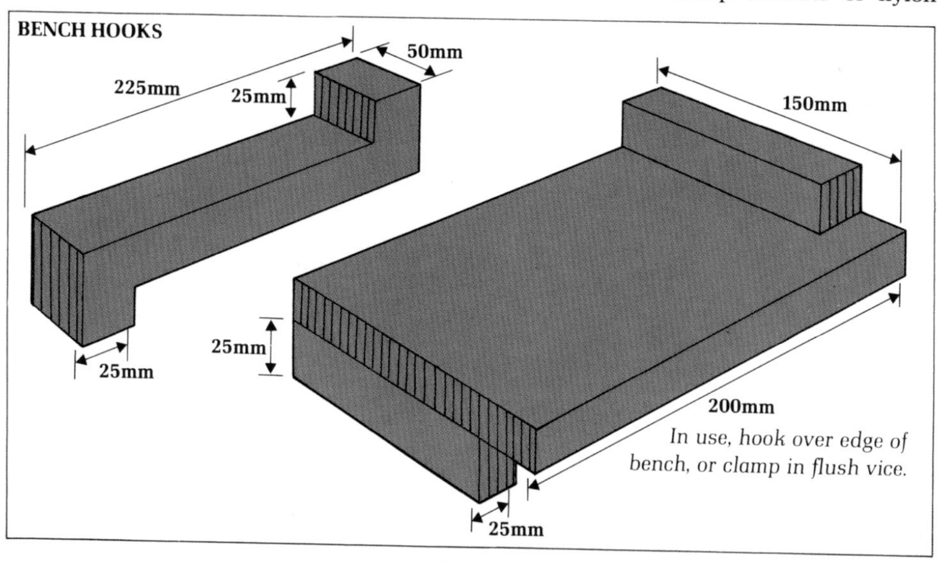

In use, hook over edge of bench, or clamp in flush vice.

webbing passed through a ratchet lever device which is used to tighten it. Web cramps are useful for irregular shapes.

G-cramps

Another useful tool is the G-cramp. This, as its name suggests, is in the shape of a letter G. It has a long, threaded jaw with a swivel head to grip at most angles. These cramps are made in a wide range of sizes from about 75mm to 300mm or more. Again, two would be useful, but four or more of various sizes would be the ideal.

Mitre cramps

One specialised type of cramp which is useful if you are making picture frames, or similar constructions, is the mitre cramp. This metal corner device has two screw-operated jaws which hold both pieces of wood firmly together. It has the advantage over the frame cramp that it may have a saw guide incorporated so that the mitres

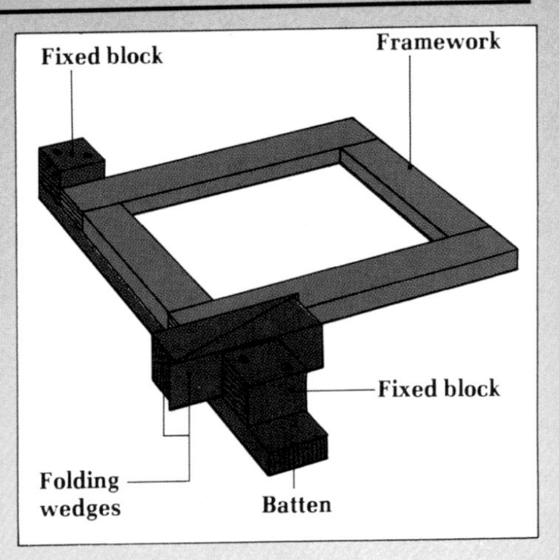

Above *Simple cramp consisting of two blocks nailed to a batten, and folding wedges to apply the pressure.*

Above right *When cramping thin wood with sash cramps, clamp battens above and below to prevent buckling.*

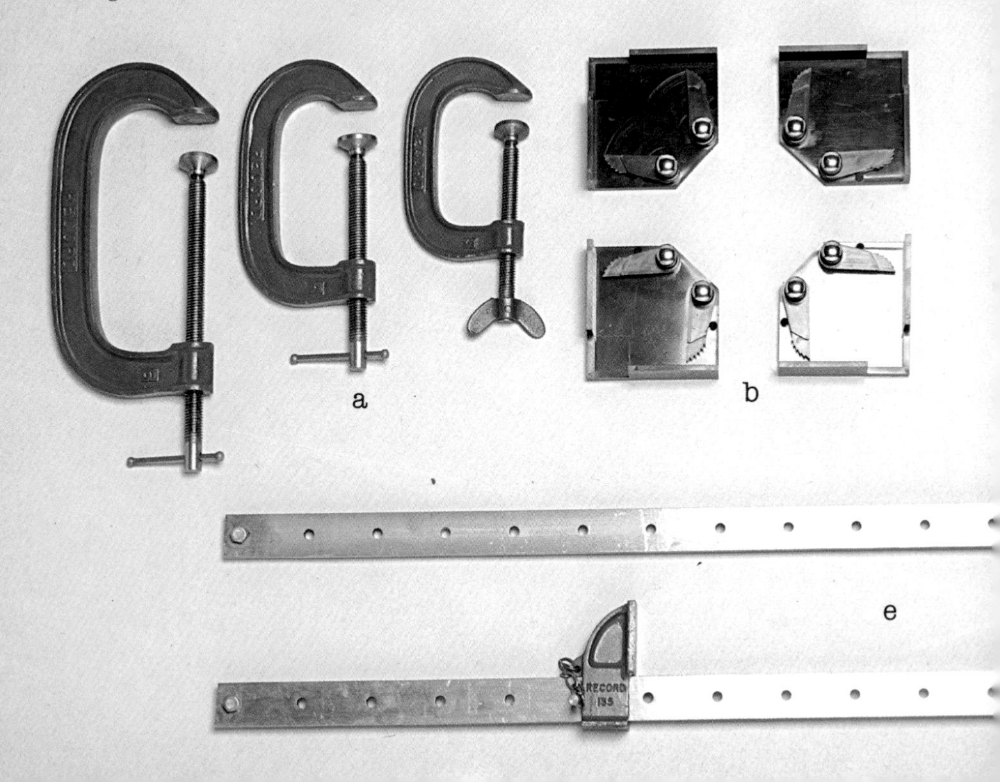

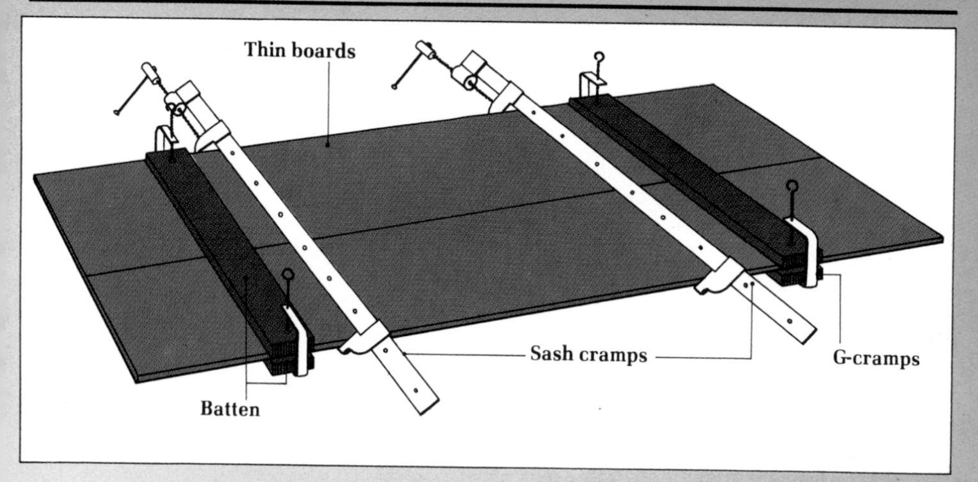

Below *Selection of cramps for general-purpose and specialist uses:*
a *Three sizes of G-cramp.*
b *Set of mitre (picture-frame) cramps.*
c *Mitre-cutting cramp.*
d *Web cramp.*
e *Pair of sash cramps.*

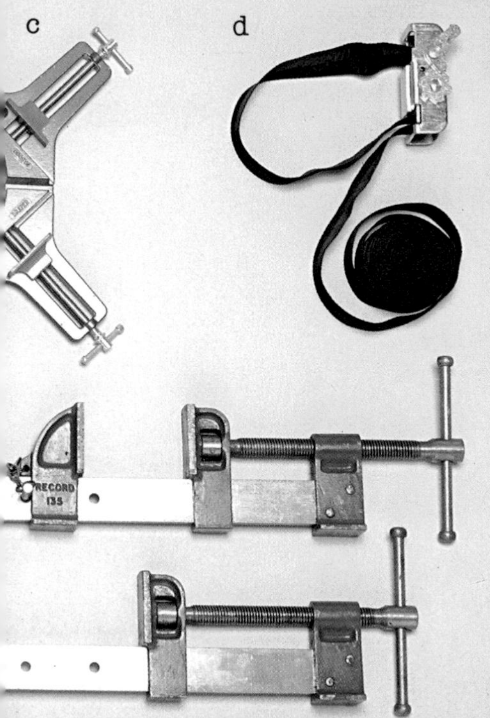

can be cut accurately while they are held in the cramp. The disadvantage is that only one corner is held at a time.

In addition to these traditional cramping devices there are other patent cramps which are designed to perform two or three different cramping actions. There are also portable workbenches which incorporate a cramping mechanism.

You can make a cramp by screwing a block to each end of a length of wood and fitting the framework to be cramped between the blocks. Then drive folding wedges between the framework and the blocks at one end.

Using cramps

Whenever cramps are used there is the danger that the edges of the timber will be marked by the pressure of the jaw, so a piece of waste wood should be placed between the jaw and the workpiece.

To overcome the tendency for frames to bend or twist under pressure, you should fit a cramp both on the top and underneath the workpiece so that they pull against each other. Frames can also be squared up by cramps. When one diagonal of the frame is longer than the other, the cramps are moved so that they are angled the same way as the frame is leaning. When the cramps are tightened it will have the effect of pulling the frame into the square position.

The start of any project, after the initial measuring up, is the setting out of the various parts. This must be done accurately if the completed work is to be satisfactory.

A wooden rule, either folding or straight, a chisel-pointed pencil, a setting-out (or marking) knife and a square are the essential tools, but others will be needed for special joints in timber.

Use the rule on its edge for greater accuracy and use the knife in place of the pencil where possible. There are in fact only five places where the pencil must be used: all rough measuring; where a cut line would be seen on the surface of the wood; for most curves; for lines which are at an angle to the grain of the wood, because a knife would tend to follow the grain; when marking out chamfers, because the cut would show as a damaged edge.

The first task on setting out is to determine which is the best side of the wood for use as the face and this should be marked so that it is easily recognisable. One edge must also be chosen as the face edge and all setting out is done from these surfaces.

Setting out dowel joints

Dowel joints, which are used a great deal when working on chipboards and blockboards, are simply marked in pencil for position. The dowel holes can be drilled by means of a jig, of which there are two or three types, providing that they are at the ends of the boards. When dowel holes are needed in the middle of the board a different approach is necessary.

Accurate measurement will provide the position of one set of holes, but to get the other set in exactly the right position to match them it is necessary to use dowel marking pins. These are simply two, short, sharp metal points separated by a ridge, like a washer. One point is pressed into

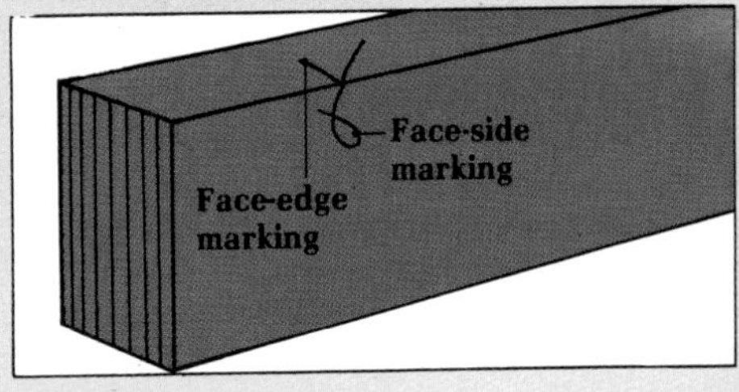

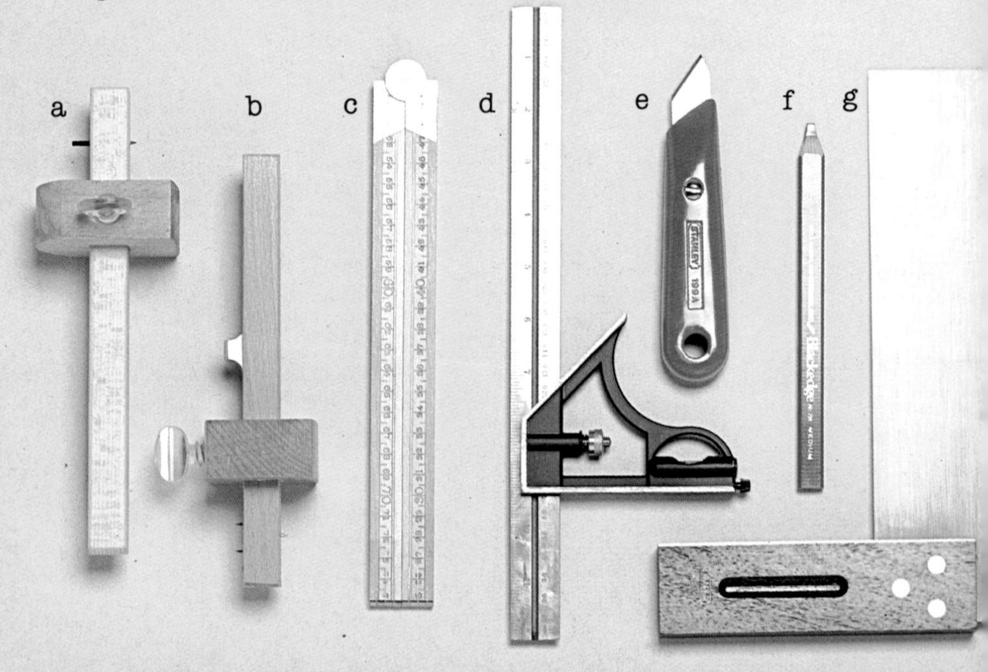

each marked position, then the second piece of board is placed on the points in the exact position that it is required. The points mark this piece so that when they are removed the holes can be drilled, using a vertical drill-stand, and will match each other perfectly.

Marking gauges

For accurate setting out of joints a gauge should be used where possible. A marking gauge with one spur will do for most setting out when depths of cut are being marked for halved joints and for housings and similar joints. This gauge can also be used for marking mortise and tenon joints, but because it has only one spur, when all the joints have been marked with one groove, the gauge has to be reset to mark the other side of the mortise and tenon.

Marking out these joints for furniture where there are a lot of tenons is much easier and far more accurate if a mortise gauge, which has two spurs, is used. With this gauge the two spurs are set at the mortise width first, usually by placing the chisel to be used between the two points and adjusting them using the screw at the end of the stem. When they are correctly positioned, the stock is set on the stem so that the mortise will be marked in the required position which is usually in the centre of the wood. Whichever gauge is used, it is important that the marking is done with the stock against one face surface of the wood.

Setting out batches

Where a number of pieces of timber, for the legs of a table, for example, are to be set out the same, they should be cramped together to hold them while the edges are marked. These marks are later squared round the sides of the wood when they have been separated again. When pieces have to be set out as pairs they are placed with their face sides together and their face edges upwards and set out as before.

Where a large number of pieces of timber are required to be set out in the same way, it is often best to set out a pattern piece and cramp it to a small number of pieces. Then set them out in batches rather than try to set them all out at once. If there are complicated joints to make, they can be drawn out full size on a length of timber or plywood so that the piece for the pattern can be laid on the drawing and the joint positions marked directly on to it.

Whatever you are making, the setting out is the most important part of the job; if you don't set out accurately you cannot expect a well-fitting final product.

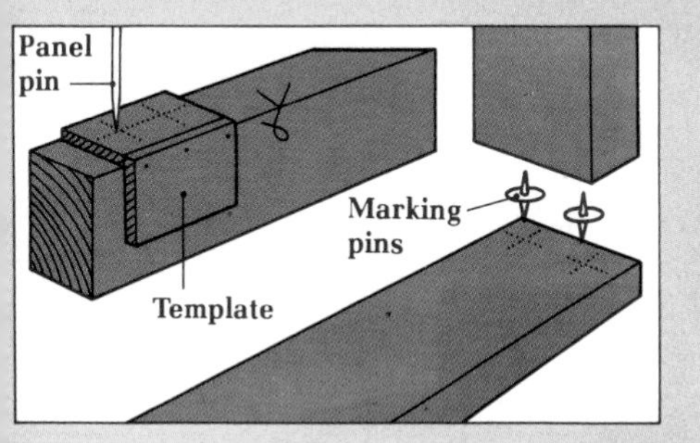

Above *setting out dowel joints using template (top) and marking pins.*

Left *Setting-out tools:*
a *Marking gauge.*
b *Mortise gauge.*
c *Folding rule.*
d *Combination square.*
e *Marking knife.*
f *Carpenter's pencil.*
g *Try-square.*

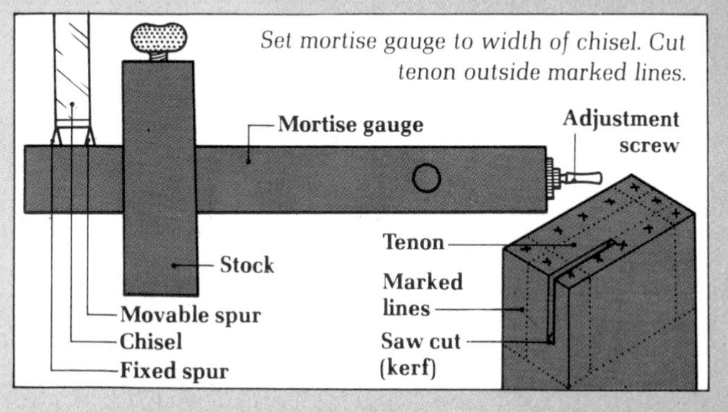

Set mortise gauge to width of chisel. Cut tenon outside marked lines.

When marking out joints, as well as when cutting them, accuracy can often be aided by using special tools or jigs. Some of these aids can be bought, others you can either buy or make for yourself.

Thumb gauge
A dowelling jig has to be bought and there are various patent types to choose from, but a little gadget like a thumb gauge is easily made and is very useful. It is simply a block of wood about 50mm square and 25mm thick. A small rebate is cut in one end so that the block can be used with a pencil to make parallel lines for chamfers or rebates. If you cut a different size rebate at the opposite end of the block you will have double value from your gauge.

Mitre-box
A mitre-box is essential for cutting accurate mitres and you can buy one or make your own to whatever size you require. A box with low sides is easier to use for small mouldings than a large box, which is intended for cutting large mouldings. By making your own you can have as many and as varied sizes as you want.

Screw the side pieces of the box on to the base and mark the 45 degree mitres using a combination mitre-square. Square the lines down the sides of the box and cut carefully down them using a tenon saw. As well as mitre guides, you can make a 90 degree cut in the box so that you can cut the ends of timber square.

Pinch rod
When you are fitting shelving in an alcove, a pinch rod is very useful for measuring between the walls. It is made from two lengths of thin batten about 25mm wide and 6mm thick. Make a brass clip or strap to fit over the battens at the ends where they overlap. Screw one to the sides of the bottom batten and the other to the sides of the top batten. This enables the two battens to be extended to the length required. If you make a neat fit in the metal straps you will be able to hold the battens in the extended

a *Mitre box for 45 and 90 degree cuts.*
b *Plunging router and bits, a powered version of the router plane.*
c *Diagonal gauge.*
d *Shooting board.*
e *Thumb gauge.*
f *Hinge-recess gauge.*
g *Dovetail template.*
h *Pinch rod.*

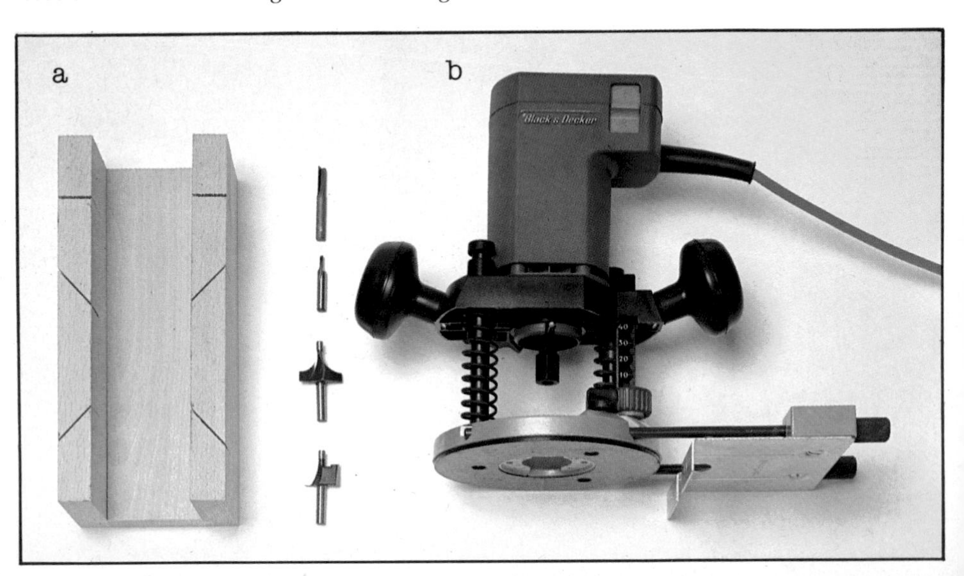

position until you have marked the timber. If the battens are a loose fit you will have to hold them with a small cramp.

Hinge-recess gauge
A countersunk-head screw driven into a 50mm square block of wood makes an ideal gauge for marking the depth of recesses for metal plates such as the leaves of hinges. Simply drive in the screw until the head is level with the plate and the head will cut a neat line in the edge of the wood.

Diagonal gauge
A thin batten pointed at one end like a chisel can be used to measure the diagonals of a frame to ensure that it is square. Push the pointed end into one corner of the frame and make a mark on the batten where the opposite corner comes. Then do the same from the remaining corners. If you put a third mark exactly halfway between the two corner marks you will have the point at which both corners will be square.

Shooting board
Mitre shooting boards can be made as well as bought. They are made from wide boards with a 50mm wide and 9mm deep rebate on one side. Two pieces of wood are then mitred together like the corner of a picture frame, and these are screwed to the base board with the point exactly on the edge of the rebate. In use, a plane is laid on its side and slid along the rebate while the mitre is held or cramped against the mitred battens on the base board. In this way, using a very sharp and finely set plane, the mitre can be smoothed to an accurate angle.

Dovetail template
Made from thin sheet aluminium or steel, this tool should be cut to the correct angle for hardwoods or softwoods.

Router plane
If you are going to make a lot of housing joints for fitting shelves in cupboards, it is important that the housings are of an even depth, so a router plane is useful. This little plane has its cutter end at right angles to the body of the blade so that it cuts parallel with the surface of the board.

Mitre-cutting cramp
This specialised tool would only be needed by those who are cutting a lot of mitres such as for picture frames. It enables the moulding to be clamped while it is cut and incorporates 45 and 90 degree saw-guides.

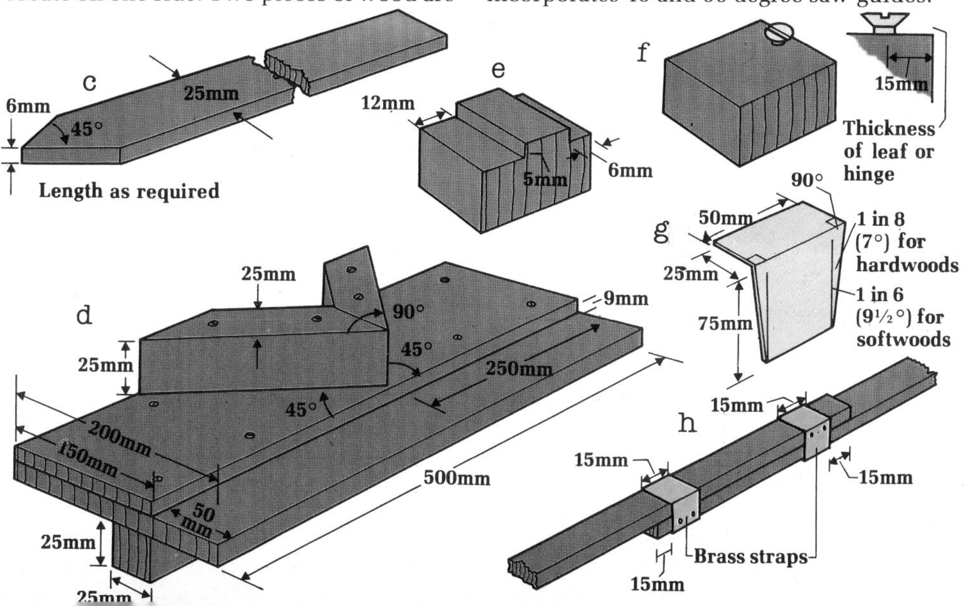

It may be thought that the use of modern resin adhesives precludes the need for traditional woodworking joints. However, there are many cases where joints such as the mortise and tenon, used for table and chair legs, are still necessary, although their strength is enhanced by a resin bond.

Mortise and tenon joint
This is used for framed-up constructions including door and cupboard frames, chairs and tables. It gives the strongest connection and the best-looking finish.

The setting out and the tools required have been described. You also need to know the correct size for the joint. A tenon should be one third the thickness of the wood and for a rigid joint it should have a shoulder all round it. If a mortise and tenon is needed at the end of a piece of wood, such as when jointing the top rails of a table to the legs, then to avoid the mortise being an open-ended slot the tenon is made only two thirds the width of the rail and a stub called a haunch is made for one third of the width. An extra 25mm of timber can be allowed on the length of the leg to avoid it splitting when the joint is assembled. This is, of course, cut off later when the glue has set.

Dowel joint
Splitting is a problem with the dowel joint, which is used as an alternative to the tenon. The dowel joint can never be as strong as the mortise and tenon joint because the size of the dowels is limited. There is no point in increasing the diameter of a dowel as this will seriously weaken the rail. A rail 25mm thick would support a dowel or tenon about 8mm thick and leave about 8mm each side to support the dowel or to form a shoulder to the tenon. If the strength of the joint needed to be increased for any reason, the tenon could be nearly doubled in thickness to 15mm, still leaving a shoulder of 5mm at each side which would provide stability to the joint. If a dowel was increased to this size there would only be 5mm at each side of the hole to support the dowel and this could prove insufficient.

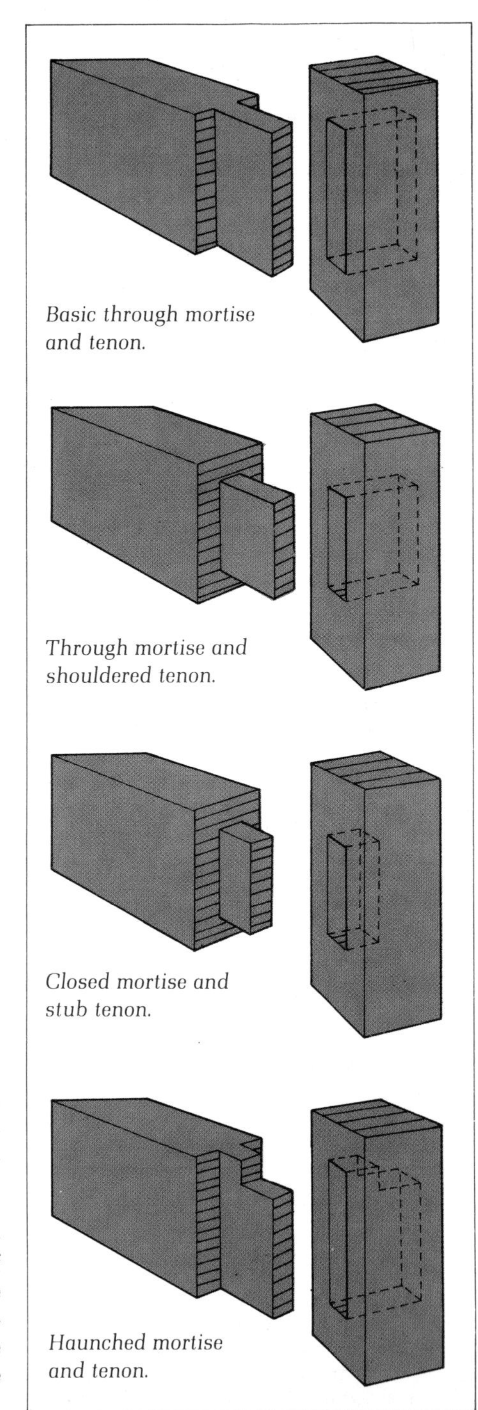

Basic through mortise and tenon.

Through mortise and shouldered tenon.

Closed mortise and stub tenon.

Haunched mortise and tenon.

This is one of the reasons for failure of the dowel joints between chair legs and rails.

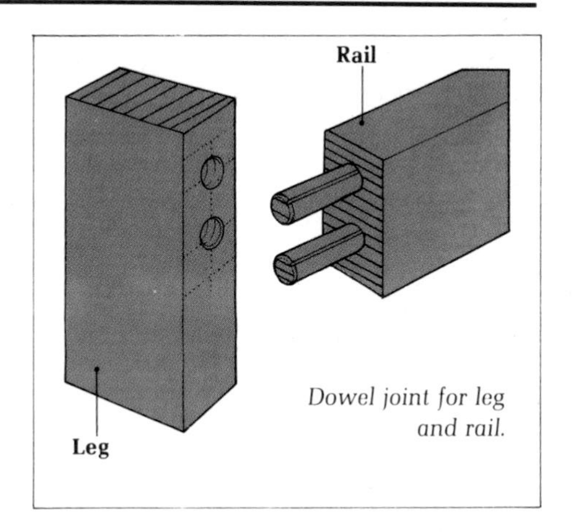

Dowel joint for leg and rail.

Halved and lapped joint
Another strong alternative to the mortise and tenon joint is the halved and lapped joint. Usually just called a halving joint, it needs no special jig to set it out. It can be marked out using a marking gauge and cut with a fine-toothed saw, preferably a tenon saw. As its name suggests it is made by simply cutting away half the timber of each piece of wood so that the two pieces will fit together flush with each other. The joint is made in the same way even if the two pieces of wood are of different thickness. In this case the gauge is set to half the thickness of the thinner piece and both timbers are marked from the face side. The face side is left intact on the thinner piece and the back is removed. The opposite is done to the thicker piece where the joint is cut on the face side to a depth of half the thinner piece. The joint can then be assembled with the two face sides being flush with each other.

The joint is both glued and screwed. Where possible the screws are inserted from the back of the joint so that they do not show when the project is completed.

Apart from corners being halved, the joint can also be used where timbers cross each other. Such places are the diagonal cross-rails at the bottom of table legs and any form of diagonal bracing.

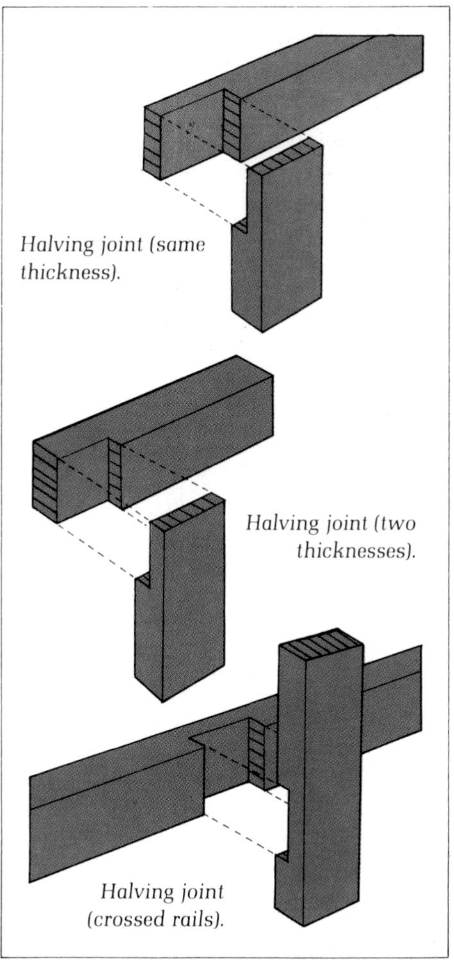
Halving joint (same thickness).

Halving joint (two thicknesses).

Halving joint (crossed rails).

Housing joint
The traditional joint for shelves which are fitted into cupboards and bookcases is the housing joint. This is a groove cut into the side member and it must not be deeper than one third the thickness of the side. In solid timber the depth need only be a quarter of the thickness. In order to prevent the joint showing, the housing can be stopped about 25mm from the front edge.

Housing joints are not really suitable for chipboard because the boards are not very thick and the edges of the groove are prone to breaking away. A chipboard shelf, even though it is not suitable for heavy weights, needs better support, and this can be provided by a thin strip, or batten, of timber glued and pinned inside the cupboard.

Butt joint

Simple butt-jointed constructions, which are made by cutting the ends of boards exactly square both in width and thickness, are suitable for solid timber which can be glued and nailed or screwed at the joint. Chipboard would have to be fixed by means of a corner block.

Butt-jointing of narrow boards to make up a wide board such as a table top is done by planing the edges perfectly straight and square in thickness and then gluing the edges before cramping the boards together. They may later be given additional support by screwing battens to the underside. Again, this type of jointing process is not suitable for chipboard construction.

Reinforcing butt joints

Although they are used as a substitute for a mortise and tenon, dowels should really only be used to strengthen ordinary butt-joints. A minimum of two dowels must be used in each joint to avoid any tendency to twist. A groove or saw cut should be made along the side of the dowel to allow the air trapped at the bottom of the hole to escape; excess glue will also be released this way. The holes in each piece of timber should be slightly countersunk to make it easier to remove excess glue cleanly. Chamfering the ends of the dowels will enable them to enter the holes more easily.

Mitred joint

This is usually made in mouldings and is secured by small nails or pins. Where extra strength is needed the joint can be reinforced by veneer keys, glued and driven into

Above right *Simple horizontal, bevelled horizontal and bevelled angled notches.*

Right *Simple bird's mouth and recessed bird's mouth joints.*

Far right *Corner and through bridle joints and box joint.*

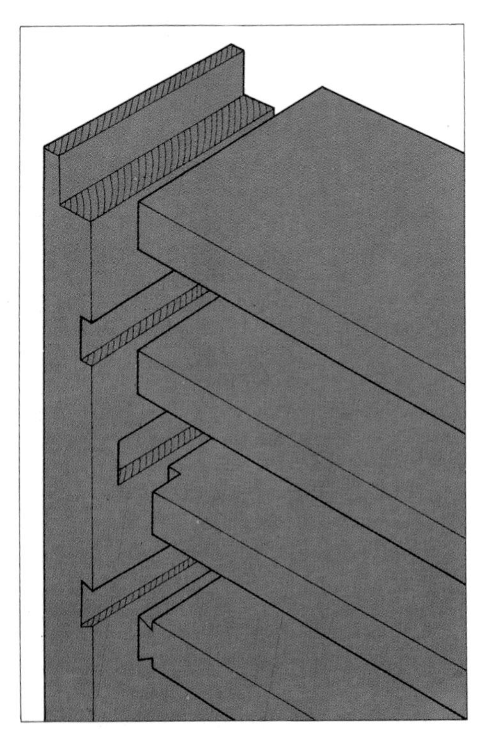

Top to bottom *Rebate; Through housing; Stopped housing; Dovetailed housing.*

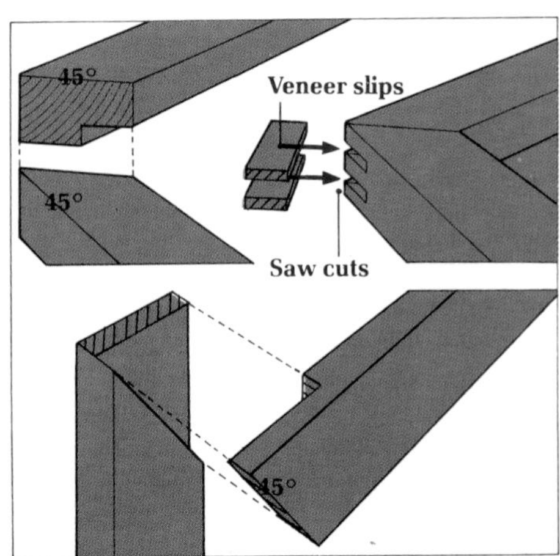

Top *Mitred joint and reinforcement*
Bottom *Mitred halving joint.*

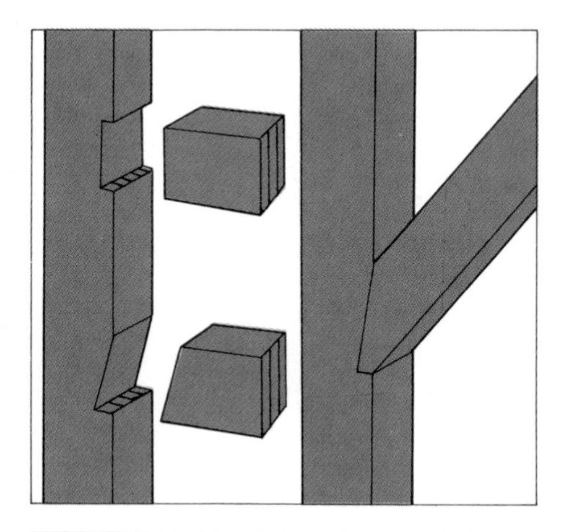

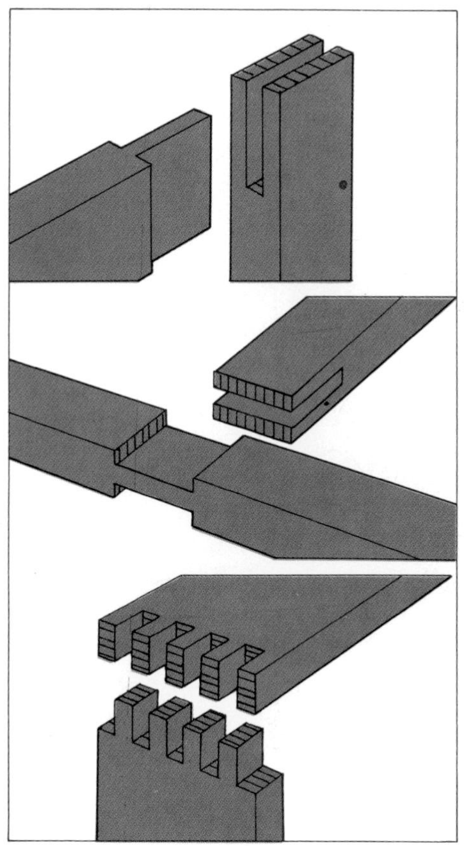

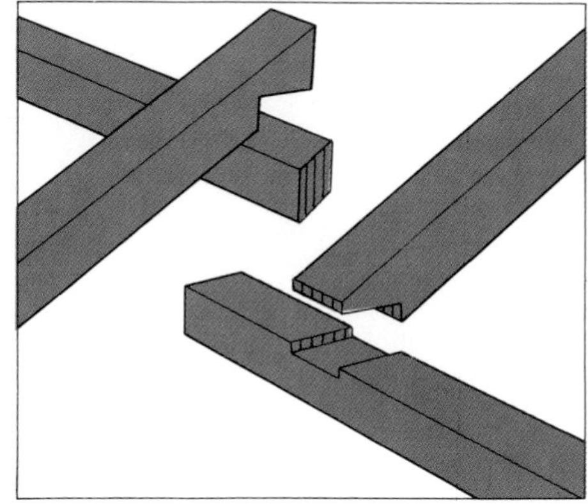

saw cuts across the point of the mitre. This is sometimes used for picture-frames.

Notching

Another simple carpentry joint is notching, which is cutting a small recess or housing in one piece of timber so that another piece can be secured in it. This is a method often used for securing the shelf-bearers where racks of shelving are being made.

Bird's mouth joint

Where an angled timber meets a horizontal timber, a bird's mouth joint is used. In this joint, a cut is made to fit the vertical face of the horizontal piece and another cut made to match the top horizontal face. The result is a right-angled v-shaped joint of a type much used in roofing work.

Bridle joint

Another carpentry joint, used in rough woodwork instead of the mortise and tenon, is the bridle joint. In this construction, which is set out like a mortise and tenon, instead of cutting a hole for the tenon, the sides of the mortise are cut away leaving the centre solid. The end of the mating piece of timber then has what would have been the tenon cut out leaving the two sides as a forked joint which will fit over the centre web.

At a corner the bridle joint looks like a mortise and tenon which has been cut too near the end of the wood leaving one end of the mortise open to make a slot. The joint is as strong as a mortise and tenon when it is made in the middle of the rail, but when used as a corner joint it lacks stability.

The range of fixing materials and devices for timber is immense. Nails, for example, are probably the simplest of fixings available, but they come in a wide range of types and sizes, each with, if not a specific use, at least a general area where they are best employed. Most are made of iron.

Round-head wire-nails are suitable for general and outdoor work where appearance is of little importance.

Round or oval lost-head nails are often used for fixing floorboards in place of the old-fashioned cut-nail. Lost-heads do, in fact, have a slight head.

Oval wire-nails are used for the same purposes as the round-head wire-nails, but because of their oval heads they can be punched below the surface so that the small hole they make can be filled with putty or other material. They can then be painted over so that they do not show.

Panel-pins are simply small slim versions of lost-head nails and are used for fixing thin material such as panels or mouldings where, being thin, they are easily covered and hidden.

Hardboard nails are also made specially for panel fixing. They have a slightly pointed head which is supposed to allow you to drive them into hardboard without having to punch them below the surface. These nails often have square shanks and are copper finished.

Clout nails are usually short, galvanised and have large heads. They are used for fixing felt to roofs.

Nails come in all lengths from 12mm pins to 150mm constructional nails. They are made in a variety of thicknesses too, so be sure to get a type which is suitable for the job. Thick nails easily split thin wood.

Nails are also made from aluminium and copper, but there is not the same range of types and sizes. There is also a limited range of galvanised or sherardised nails.

Screws

Screws are equally prolific, although they are most freely available in a more limited range of preferred sizes. You can still get the non-preferred sizes, but you may have difficulty finding them, and they may be more expensive.

Head types

The main types of head are countersunk for all general applications, raised countersunk which are usually plated and used for fixing metal fittings, and round head which are often used for fixing flat metal and may be used in conjunction with washers. There are also pan heads which are similar to round heads but have a flatter section. Other heads are made and are often used for heavyweight coach-screw fixings.

Materials and finishes

Screws are made in steel, brass, aluminium, stainless steel or silicon bronze. There are also a number of finishes including sherardised, nickel plate, chromium plate, brass plate, bronze metal antique, dark Florentine bronze and black japanned.

The plated finishes are only suitable for interior work. For external applications the solid brass, stainless steel and sherardised screws should be used. Aluminium screws are ideal for fixing cedar as they will not

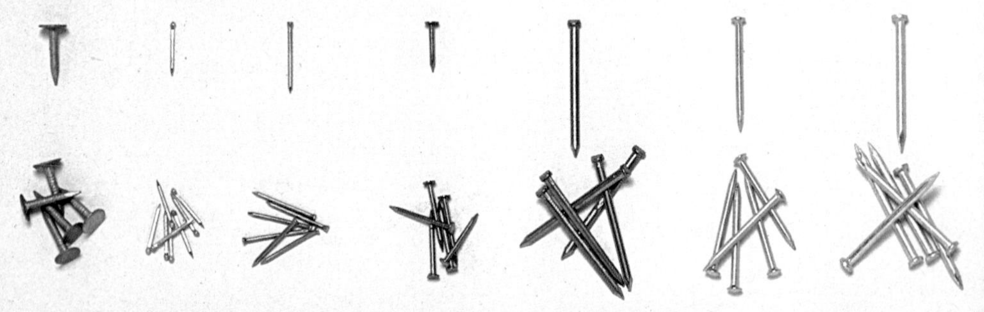

stain the surface. They are also, as are solid brass screws, suitable for fixings in kitchens and bathrooms where the damp atmosphere would soon cause ordinary steel screws to rust and stain surfaces.

Screw threads

There are two types of thread on screws. The first and most often used is the tapered thread which has a single spiral running down the shank. The other type is the chipboard screw. This has two spirals running round a parallel shank. The shorter lengths of these screws are threaded up to the head. The parallel shank reduces the tendency to split the chipboard.

Using screws

Whichever screws are being used it is always best to drill pilot holes. If these holes are the same diameter as the shank of the screw below the thread, then a full grip will be obtained without danger of the wood splitting.

The size of wood screw required for a particular application depends upon the width and thickness of the timber. The diameter of the screw should not exceed one tenth the width of the wood into which it will be inserted. You should also ensure that at least four diameters of thread, and if possible up to seven, are engaged in the wood. If the screw is fixing a piece of wood, its length should be not less than three times the thickness of the wood it passes through. If pilot holes are drilled the screws can be positioned up to 10 times the diameter from the end of the wood or five times the diameter from the edge. If pilot holes are not drilled then the screws must be kept 20 times the diameter from the end of the wood.

If the screws are lubricated with soap, tallow, beeswax, or Vaseline, if grease stains are of no importance, you will find it easier to drive them, especially into hardwoods. They can also be removed more easily at a later date if necessary. The lubricant should be applied to the point of the screw or into the pre-drilled hole. Avoid getting any lubricant on the head of the screw as the screwdriver may slip.

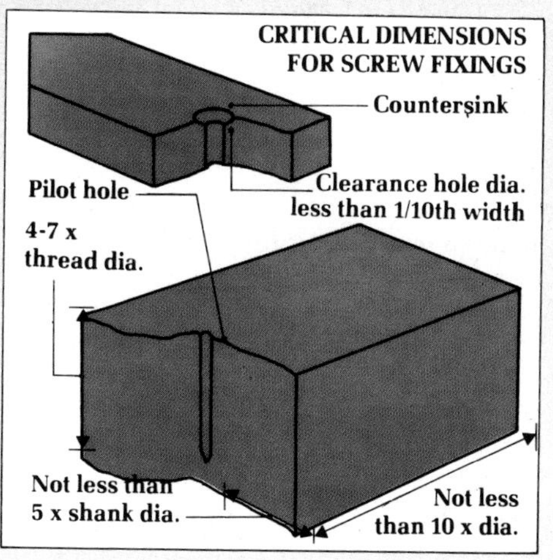

Below *(left to right)*:
Clout nail; Hardboard nail; Panel pin; Round-head wire-nail; Oval wire-nail; Batten-head nail; Masonry nail; Chrome-plated brass round-head woodscrew; Zinc-plated chipboard screw; Chrome-plated brass raised-head woodscrew; Four sizes of steel countersunk woodscrew.

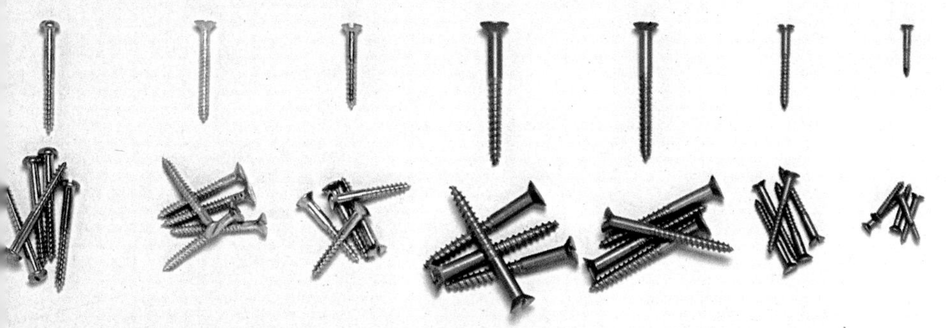

At one time woodworking required only one type of glue – Scotch glue – which was boiled in a pot and used hot. This glue, although little used today, is still available. It has only moderate strength and it is not waterproof.

Today there are various types of adhesive which you can use. Some are in tubes and are ready for immediate use and others are in powder form and have to be mixed as required. Some of the resin glues which are waterproof are also extremely strong. A carefully made joint using these glues would not, after it had set, break along the glue line; it would be the surrounding wood which would fail under stress because of its lower strength.

Woodworking adhesives

For small woodworking projects you can use most household adhesives. The most universal is the PVA type of adhesive which will bond many materials to wood as well as fixing wood to wood. It has the advantage that surplus glue can be wiped off the surface with a wet rag before it sets. The powdered glues which are mixed with water can also be wiped off with a wet rag. The PVA adhesives are not waterproof, so for joints which have to withstand water you must use one of the urea formaldehyde resin adhesives which have a powerful gap-filling capacity.

a *Contact (impact) adhesive.*
b *Resin woodworking adhesive.*
c *Two-part epoxy resin adhesive.*

Contact adhesives

When veneering in wood you can still use Scotch glue or PVA, but when fixing plastic laminates to timber you have to use one of the contact adhesives. The type which gives instant grip is suitable if there is scope for finishing operations which would hide any slight error in positioning the laminate. The alternative is to use thixotropic contact adhesives which allow a slight movement before the final pressure is applied.

Epoxy resins

For extra strength there are the epoxy resins which are two-part adhesives requiring the application of a catalyst or hardener to create the required bond. With some of these glues you mix the hardener into the solution just before it is applied to the surface. There is then a time limit on assembling the parts – though this generally gives plenty of time. With the other type, the glue is applied to one part of the joint and the hardener to the other part, then when the two are brought together the necessary reaction takes place.

Such powerful adhesives are not often required for domestic woodworking and carpentry, although they could be useful for assembling furniture and projects to be used outdoors.

Finally, is there any difference between a glue and an adhesive? According to one manufacturer "No, it depends on the size of the package – some are too small to take the word adhesive and leave space for other more important information like the maker's name."

There are two main types of construction in woodwork. One is box construction in which the project is mainly composed of sheet materials which are fastened together without being attached to a framework (see p.34). The other is frame construction in which lighter panels are supported on a framework. As this framework is made rigid by the various joints which are used in its construction it is better able to support doors and is less likely to sway with the movement of the doors.

Mortise and tenon joints

The basic joint for framed construction is the mortise and tenon. This is used because the shoulders of the tenon make the frame rigid and the tenon itself enables the members to be joined without seriously weakening the timber at that point. For cupboard frames the ends of the top and bottom rails are tenoned to fit into mortises in the sides or stiles, while for door frames and window frames the joint is made the opposite way round: the sides or jambs are tenoned to fit into mortises in the head and sill.

Halved and lapped joints

These joints can be used as a substitute for the mortise and tenon, but when this is done the appearance should be preserved by allowing the stiles of the cupboard frame to run up the face of the top and bottom rails. They cannot be used for window and door frames, but a housing joint can be made in the head and sill so that the jamb can fit into it. This joint would have to be well nailed. Doors which are to be clad with hardboard or plywood can also be made with halving joints. It is best to fix the hardboard panel to the frame using an adhesive because the nails always seem to show through the paint however careful you are in punching them down and filling the hole.

Dovetail joints

Traditional box construction uses dovetail joints as these give a pleasing appearance at the corners of the furniture. This joint is not used much today because of the amount of veneered chipboard which is used for furniture, and dovetails cannot be made successfully in chipboard. Some machine-made dovetails are used in quality furniture but they lack the decorative appearance of hand-made joints because the dovetails and the pins which separate them are all the same size.

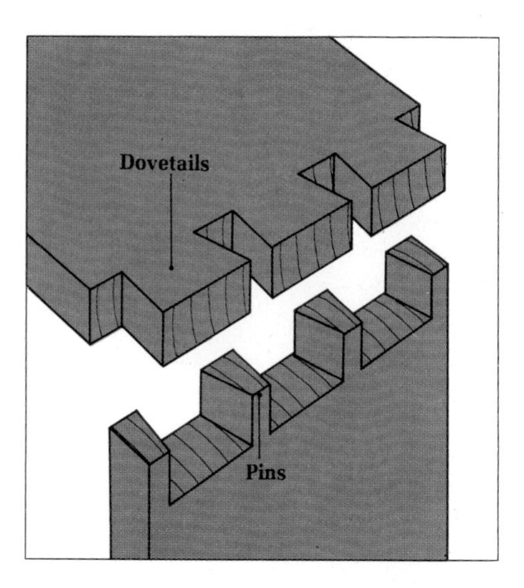

To make dovetailed bracket, cut angled notches in vertical and horizontal pieces to house bracing piece.

However, dovetails provide a joint which will resist a lateral pulling force better than most other joints, so they are especially useful when making wooden brackets. Here the dovetail is cut in the top member, and the top end of the upright member has the cut-out for the tail to fit into. If the angled member of the bracket is let into the top and upright member, then the bracket will have considerable strength.

The correct angle for the sides of the dovetail is one in eight (7 degrees) for hardwoods and one in six ($9\frac{1}{2}$ degrees) for softwoods. If the sides are made too steeply angled the pointed corners of the tail will break and if the angle is too shallow it will have little resistance and the joint will pull apart easily.

There are fittings for almost every conceivable requirement. Hinges are available in patterns to suit any cabinet. The most popular for kitchen cabinets is the concealed type which gives the door a pivoting action so that the cabinet can be fixed close into a corner. Some of these hinges incorporate a spring so that the door locks itself into the open or closed position, doing away with the need for a separate catch.

The simplest hinge for a glass door has a pivot action. A hole is drilled about 10mm deep in the top and the bottom of the cabinet and a plastic bush is inserted. Into this bush fits the pivot of the hinge, then the glass is slid into its channel where it is held in place by grub screws.

Handles must be chosen to suit the cabinet and its location. For kitchens, robust metal channels are generally used as they will stand up to heavy use. The more decorative handles used for lounge or bedroom furniture can be in wood or metal and may screw into the face of the door or drawer or, in the case of the ring-pull types in brass, they need to be carefully let into the surface to produce a flush finish.

Magnetic catches are the easiest to fix and are neat in appearance. Some catches are of the touch-latch type which open partly at a light touch allowing the door to be fully opened. Spring operated touch latches are

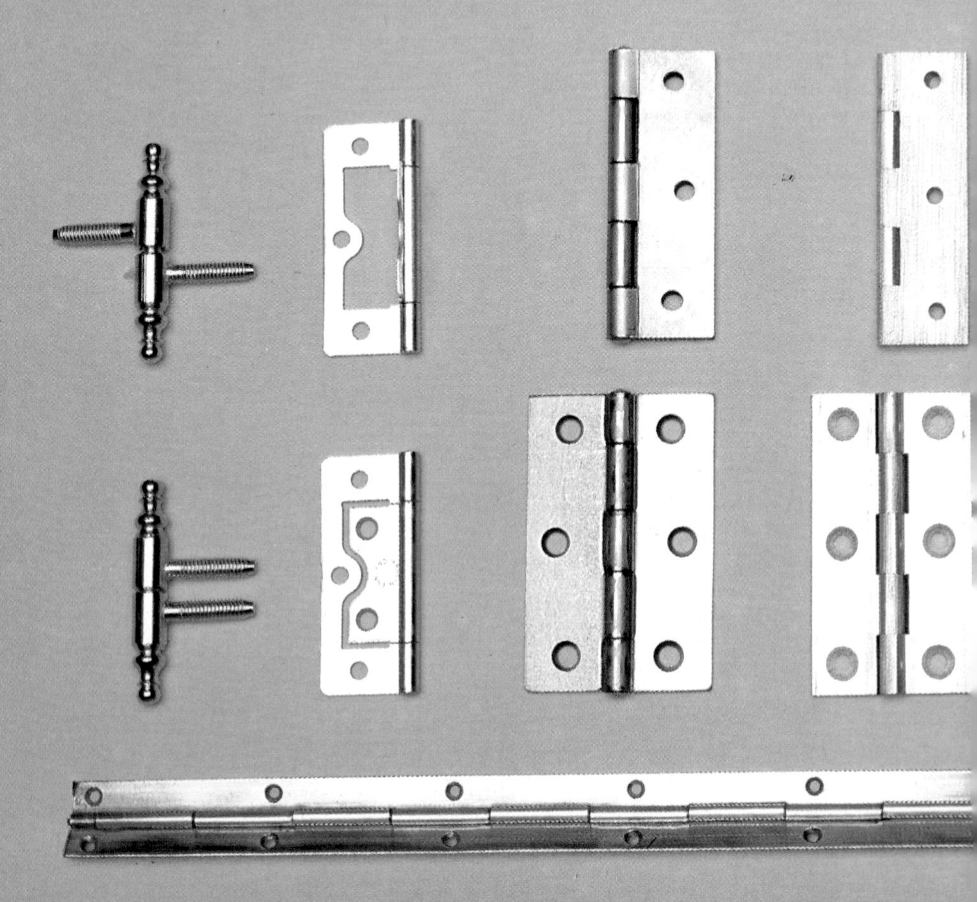

also available. In all cases fitting involves screwing the catch to the framework; recessing or drilling is not often required.

Bolts for the insides of double doors such as wardrobes and bookcases are usually in brass and are flat, not round. They may screw directly to the door or they may be let into the surface.

Adjustable shelves can be provided in bookcases by means of slotted strips screwed to the sides of the cabinet. A small clip fits into the horizontal slots to support the shelves. Four strips are required for each cabinet.

More difficult to position accurately, but less obtrusive, are shelf studs. These need carefully set out and drilled holes into which the studs are pushed. Two columns are required at each side of the cabinet.

Sliding door gear is obtainable in all sizes from simple channels to heavy overhead rollers. It is necessary to choose from the range made for the weight and type of door which you wish to hang.

Whatever type of cabinet you are preparing to build, always ensure that the fittings of the type you need are obtainable in the sizes you require before starting work. It is best to have all the fittings to hand at the outset, otherwise you could get into problems if you look for fittings only after the work is well advanced – well thought out planning is essential.

These small hinges and fittings are easy to fit and are suitable for a variety of cabinets. They include, on the opposite page (left to right): pivot hinges; flush hinges which require no recess; pressed-steel butt hinges; narrow-leaf brass butt hinges; and piano hinge (bottom). On this page: automatic touch latch (top left); concealed sprung mini hinges which need no catch (top right); mini automatic latch (bottom left); magnetic catches (bottom centre); and sliding and fixed mirror clips (bottom right).

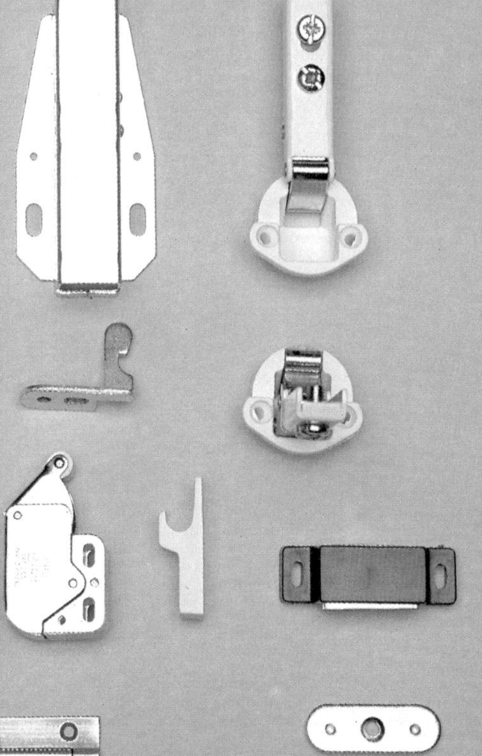

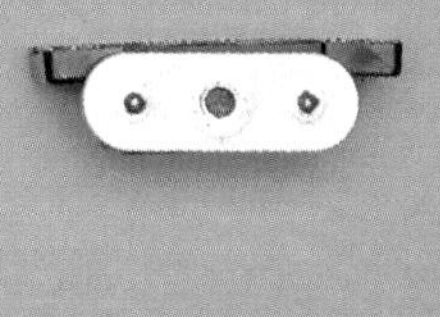

Hanging doors calls for care and neatness if the finished job is to look acceptable. The framework must be rigid if the door is to be hung within it. You can use a piano hinge screwed directly to the edge of the door or you can use butt hinges which have to be let into the edge of the door and the cabinet side. In either case, the door must be carefully fitted into the framework so that it will show about 1.5mm joint all round when it is hung. With a piano hinge you will need to make the door equal to the width of the opening less 1.5mm and the thickness of the hinge. Butt hinges need an even joint all round as they are let into the wood.

It is obvious that to make a good job takes a little extra care and if the framework is not rigid it will move slightly when the door is hung and cause it to bind at some point.

These problems can be overcome by having the door laid on the face of the framework. It will then be cut to the same size as the outer dimensions of the cabinet with no need to plane it to an accurate joint. There are two types of lay-on hinge, one that is partly let into the door and one that simply screws into place.

The lay-on hinge is screwed to the door allowing for the thickness of the cabinet side. The door is then hung with the hinge screwed flush with the front of the cabinet. This hinge can also be used for inset doors if the hinge is screwed to the side of the cabinet set back the thickness of the door and also set back from the edge of the door the thickness of one leaf of the hinge.

The other type of hinge for lay-on doors has a round boss on one leaf that has to be let into the door. A special end mill bit is necessary for boring the blind hole. An ordinary bit has the centre point too long and it will pierce the front of the door before the hole is deep enough. The best end mills, which are 26 or 35mm diameter, have built-in depth stops to ensure that the holes are milled to exactly 12.5mm deep.

Holes are 3mm from the edge of the door which makes 20.5mm to the centre of the hole for 35mm diameter and 16mm from the edge for the 26mm diameter holes. The boss is set into the hole and screwed into place. Then the door is offered up and the cabinet marked so the base plates can be screwed into position. The door is hung by screwing the leaf of the hinge on to the plate where the correct closure of the door can be adjusted by two screws.

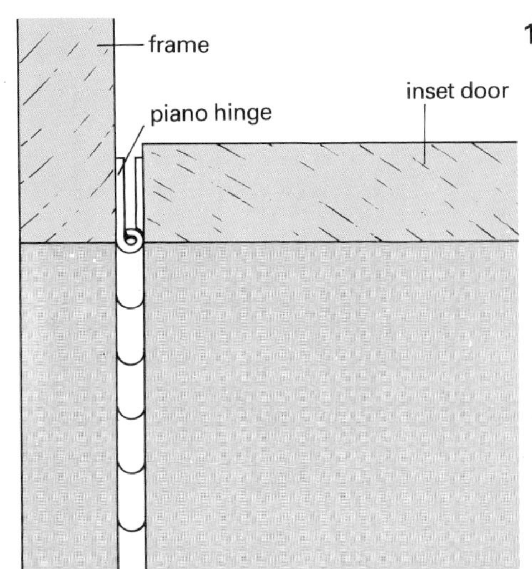

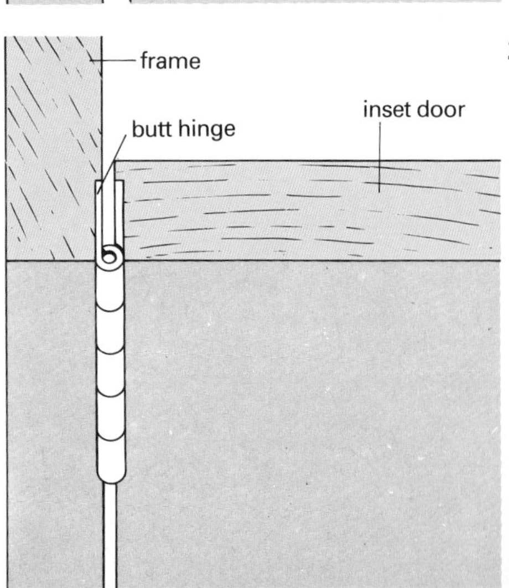

Two of the main types of hinge are illustrated here: piano and butt hinges which are suitable for the doors of furnishing cabinets, and concealed hinges which are more suitable for the doors of kitchen cabinets.
1 *Piano hinges are useful for chipboard doors because they cover the raw edge and offer a large number of screw positions.*
2 *Butt hinges can be let into the frame and the door.*
3 *Both leaves of the butt hinge can also be let into the door.*
4 *The lay-on hinge needs no recesses.*
5 *The concealed hinge needs a circular recess in the door.*
6 *The hinge and the drill, or mill as it is called. Both these last two hinges allow doors to pivot close to a wall.*

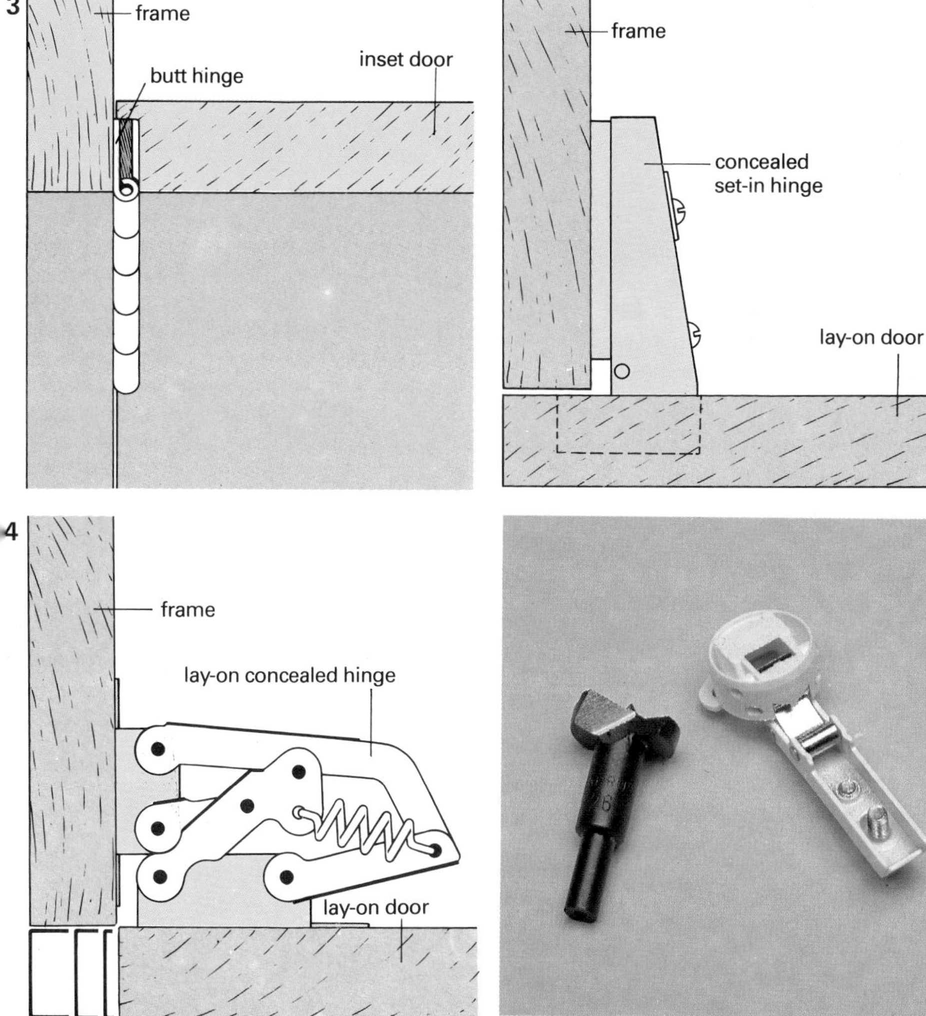

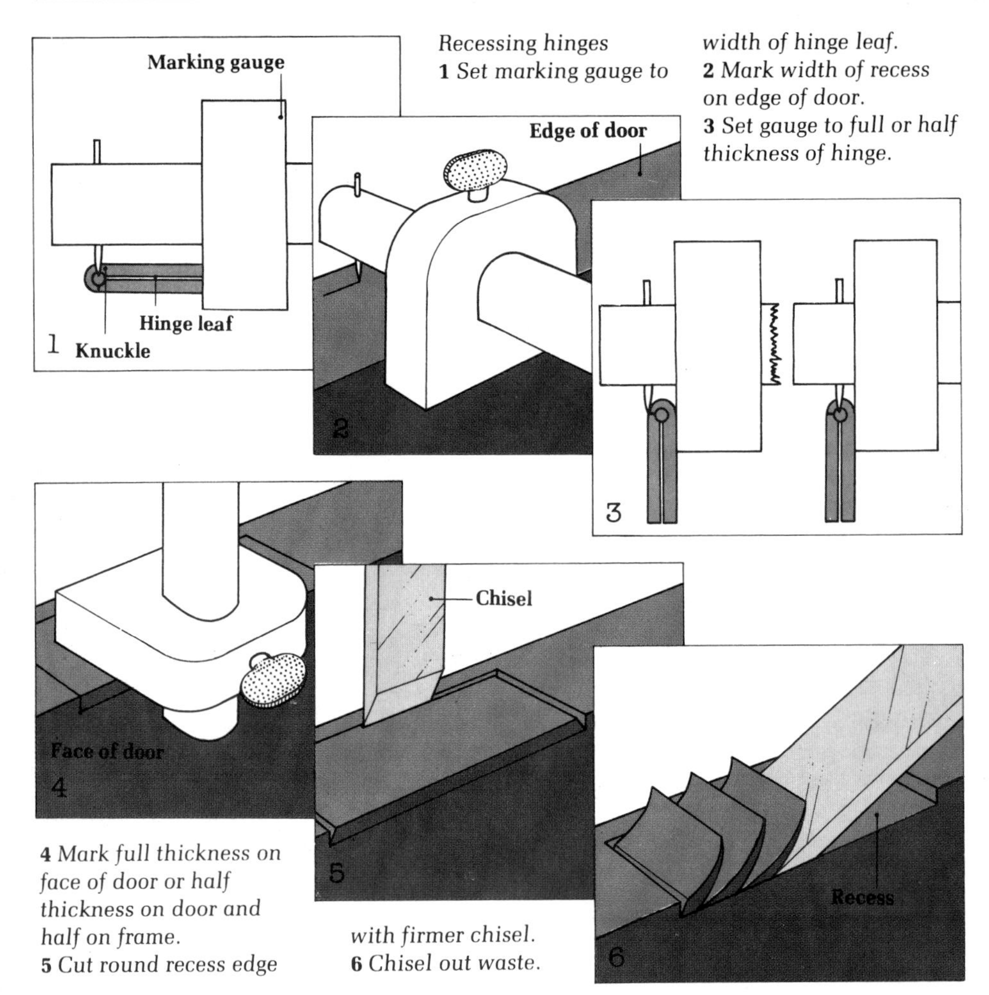

Recessing hinges
1 *Set marking gauge to width of hinge leaf.*
2 *Mark width of recess on edge of door.*
3 *Set gauge to full or half thickness of hinge.*
4 *Mark full thickness on face of door or half thickness on door and half on frame.*
5 *Cut round recess edge with firmer chisel.*
6 *Chisel out waste.*

Doors for solid-wood cabinets made in the traditional manner are set inside the framework and not laid on the face of the cabinet as is frequently done with modern furniture made from veneered chipboard. Carpenters usually let one leaf of the hinge into the frame and the other leaf into the edge of the door, but cabinet-makers often let the whole of the hinge into the edge of the door so that the hinge does not break the neat joint line. The hinge position should be just clear of the mortise and tenon or other joint which may have been used so that the fixing screws grip into solid timber.

Fit the hinge to the door first. As this is cabinet work, there will not be a number of coats of paint to be applied. The wood will only be polished, so the door can be planed to a neat fit with only a minimum of joint allowed. When the hinge is in place, the door is inserted into the cabinet and the position of the hinges is marked on the frame. Recesses are made for the hinge leaf or alternatively the hinge is simply screwed to the frame.

Be careful to use the correct size of screw. If the heads are too big and do not fit into the countersunk holes in the hinge they will bind on each other when the door is closed. Should too little be cut out of the hinge recesses, the door will strike the cabinet and not close. If too much is cut out of the recesses, the door will bind on the hanging side. This will also occur if the door edge is not planed square but bevelled slightly to the front. It is best to err slightly on the side of taking off a shaving too much on the inner edge of the door as this will prevent the door binding when it is closed.

Four-legged constructions

Where constructions such as chairs and tables are being made, the mortises for the rails, which are at right angles to each other, meet inside the leg. This means that the tenons must be mitred at their ends to give the maximum length to each.

Tops of solid-wood tables must be fixed by means of slotted metal plates or by wooden buttons which engage in grooves in the inside face of the top rail. This is to enable the wide wooden top to move with changes in atmospheric conditions. Solid tops which are battened on the underside must also be allowed to move. This is done by not gluing the battens, and making slots for the screws instead of tightly fitting holes. Failure to provide for the movement of natural timber tops could cause serious splitting in a warm atmosphere.

Side panels fitted into grooves in the framework must also be allowed to move and therefore they are not made a tight fit and they are not glued or pinned. Chipboard and laminated boards do not suffer this and can be securely fixed.

The legs and rails of chairs and tables, for example, must be cramped up carefully to avoid the whole article becoming twisted. Sight across the construction, viewing one rail against the same one on the opposite side, and you will be able to see whether they are in line or whether they are sloping in opposite directions. If they are not all in line, the piece of furniture will rock or wobble on its feet.

To make a wobbly chair or table stand firmly, wedge it level on a flat, level surface. Take a strip of wood about 3mm thicker than the amount the leg is short and mark a cutting line against the top of it on each leg. Never cut one leg to suit the others.

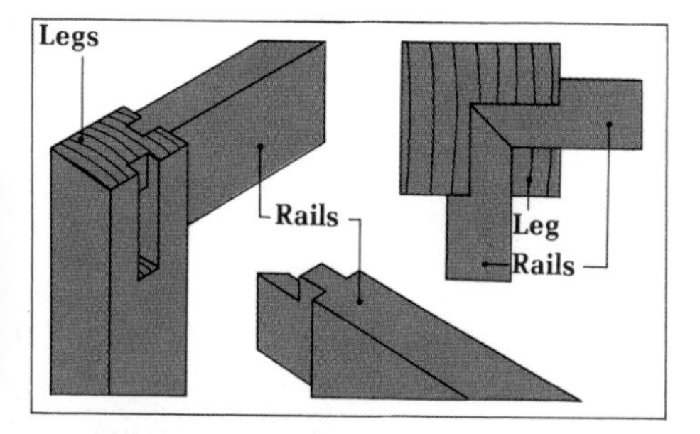

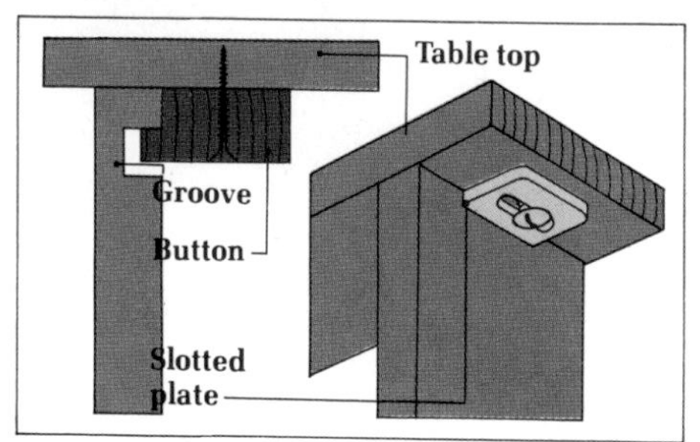

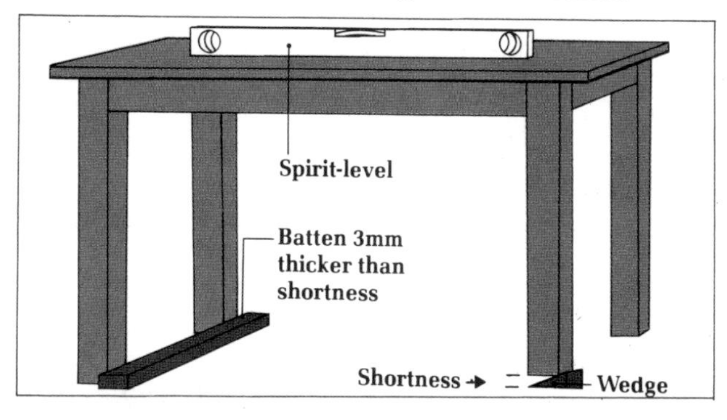

Top *Mitred haunched tenon joining rails to top of table or chair leg.*

Above *Join wooden table-top to base with wooden button (left) or slotted metal plate to allow for movement of timber.*

Right *Trim all four legs to level a wobbly table.*

Working with man-made boards requires different techniques to conventional joinery and involves very few cut joints.

Chipboard

Much modern furniture is basically boxes made from veneered chipboard. Construction is often of the knock-down type so that the units can be taken apart and fitted into a cardboard box which will go in a car boot or on a roof-rack. Any permanently constructed units are generally joined by means of dowel joints. There are two main requirements for making furniture like this and they are the ability to cut a straight, square line and to drill neat, perpendicular holes. This latter requirement is easily met with a dowelling jig or vertical drill-stand.

This type of furniture is ideal for the beginner to make, as the boards can be bought cut to length and plastic blocks can be used for the joints and for supporting shelves. The blocks are simply screwed into place using double-threaded chipboard screws, enabling you to produce the furniture quickly. Doors can be of the lay-on type so that there will be no need to plane them to fit into an opening. Hinges can be of the concealed lay-on type which need no cut-outs and are just screwed into place.

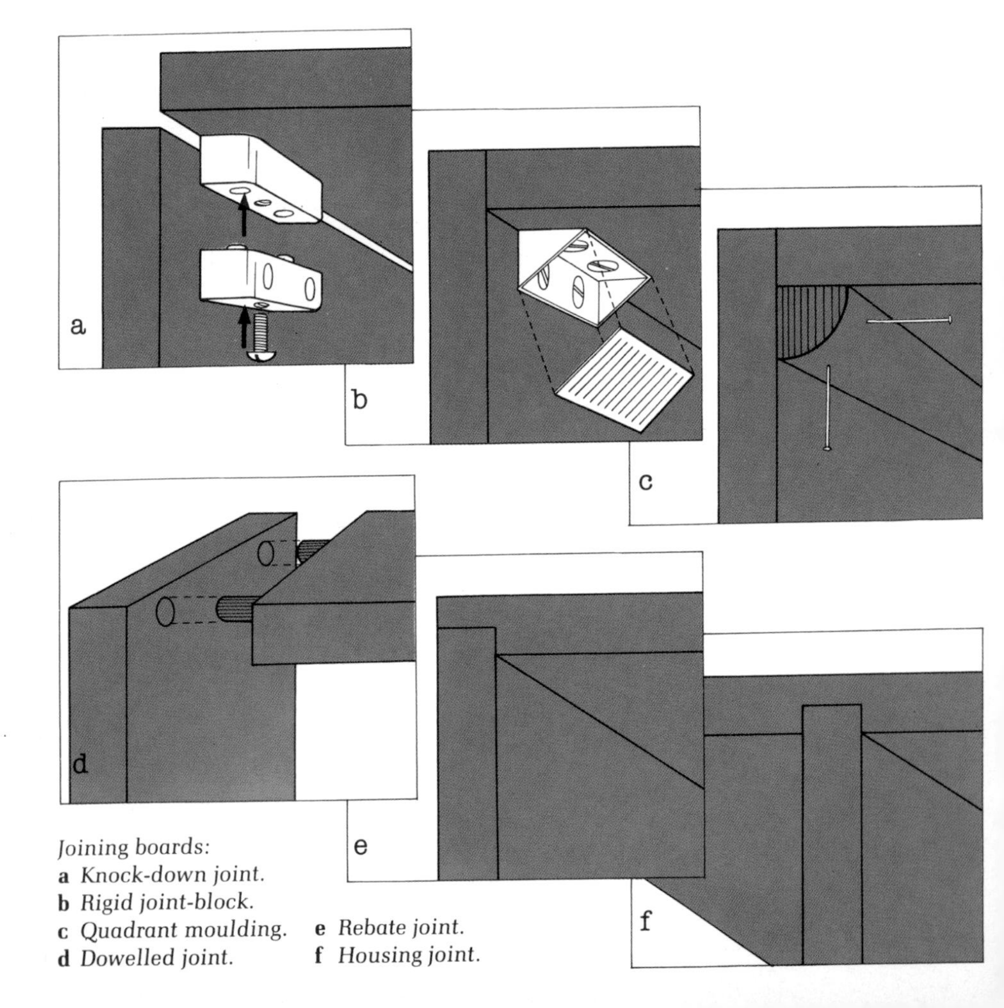

Joining boards:
a *Knock-down joint.*
b *Rigid joint-block.*
c *Quadrant moulding.*
d *Dowelled joint.*
e *Rebate joint.*
f *Housing joint.*

Laminated boards
Working in blockboard and pine-board follows a more traditional method, as some woodworking joints can be employed. This kind of board will accept a mortise and tenon joint as well as a notch or a rebate in the edge. The boards will also accept screws in the edge which is not a good practice when using chipboard. To some extent you can make housings into the board so that shelves can be fitted.

Hardboard
This can be used for making doors and for cladding other types of framing, but it must be conditioned first by damping the back of the board, using about 1 litre of water to a 2440 × 1220mm sheet. The wet sheets should be placed back to back for 48 hours before use. This process enables the board to stretch so that when it is fixed and dries to the moisture content dictated by the conditions in which it is being kept, it will tighten up and will not buckle. It would do so if it were perfectly dry when it was fixed to the framework and was then taken into a more moist atmosphere. The treatment is not necessary for free-moving panels.

Cutting boards
Sheet materials, especially thin boards like hardboard and plywood, are cut most easily if they are supported along the whole length of the cutting line on both sides of the cut. Lack of support allows the sheet to sag or bend and jam the saw.

Rigidity
Because there is no basic framework to hold the whole construction rigid, stability depends on the hardboard back of the unit being fixed securely to the sides and rails or shelves. In addition, any plinth or top rail will help to prevent the front of the unit swaying. This can prevent sliding doors from meeting the sides properly and it will make hinged doors swing open or closed.

Right *Maximum span between supports for various types and thicknesses of shelf.*

Finishing
The plywood facing of blockboard and the surface of pine-board can be stained, polished or varnished without preparation other than sanding. This also applies, of course, to veneered chipboard and plywood, but plain chipboard needs some degree of filling, or preferably veneering, before a suitable finish can be applied.

Cut edges of chipboard can be a problem, but they can be trimmed using iron-on veneer strips which are obtainable in finishes to suit the finish of the boards you are using. Careful planning will enable you to use standard sizes and therefore reduce the number of cut edges. You can also arrange for the cut edges to be at the back, where they will be covered by the hardboard back of the unit, or in a position where they will be covered by other boards.

When plain chipboard is being used and is to have a decorative laminate applied to it, it is important that a similar laminate, which will be cheaper because it is not decorative, is applied to the back as well. Without this balancing veneer, the chipboard will tend to bend. However, the balancer is not essential when the board is being used for worktops or table tops because these can be screwed securely to the frame which will restrain it. Unlike natural timber, chipboard does not need to have the freedom to move with the changes in the atmosphere.

Type of board	Thickness (mm)	Maximum span (mm)
Chipboard	12	400
	19	600
	25	760
Blockboard & Plywood	12	450
	19	810
	25	1000
Pine-board	19	810
Timber	16	500
	22	914
	25	1000

Free-standing shelves must be well made and braced if they are to be secure when carrying heavy weights. It is always best to try to have one end fixed to a wall as this will do away with the need for bracing the structure.

The size of the uprights will depend on the weight likely to be carried; an average size suitable for general storage would be 40mm square. All the uprights must be cut off to the same length and their ends squared. One of the pieces is then set out with the positions of all the bearers. Then all the uprights are put together with their ends level so that the marks can be transferred by squaring them across. This will ensure that all the members are marked exactly the same and all the bearers will be in line.

Bearers can be either screwed directly to the face of the uprights or they can be partly let in for extra strength and stability. A housing about 10mm deep would be sufficient and would not weaken the upright too much. The housing is cut with a tenon saw; extra cuts in the central waste will make chiselling it out easier. Chisel from each side of the housing

1

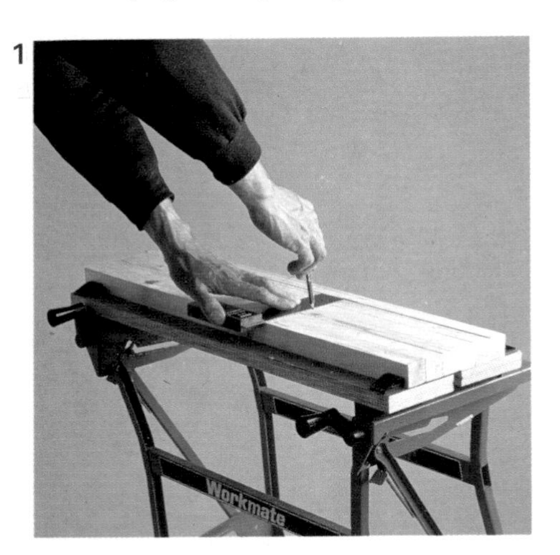

1 When making free-standing shelves all four legs are set out together.
2 The legs should be held firmly while the notches for the bearers are cut.
3 If the bearers and notches are a good fit, they will make a square frame.
4 Notching the ends of the shelves helps to make the structure rigid.
5 A brace will be necessary and it can be let into the legs and shelves for greater strength.

2

towards the middle to make a neat joint. Drill the bearers and screw them to the uprights. If neat joints have been made the ladder sections which you have made will be fairly rigid.

The shelves can be fitted either between the uprights or they can be cut to fit round the uprights to finish flush with the front edges. The latter method helps to make a more rigid structure. In either case some bracing will be required if the shelves are to support any great weight. Even if only open slat shelves are used, it is best to notch the outer slats round the uprights to give the extra stability.

At least one brace will be needed at the back. Lay a piece of timber, usually 50 × 25mm, from the top shelf down to the rear upright. Mark where it crosses the shelves and, if there are any, intermediate uprights. It should cross at least two shelves. Notch out these joints to a depth of about 10mm and screw the brace into place ensuring at the same time that the framework is square. This is done by measuring both diagonals of the structure from corner to corner, and adjusting it until they are equal.

3

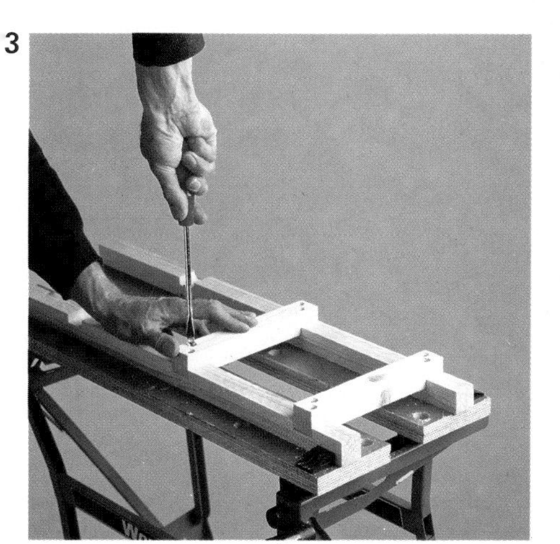

4

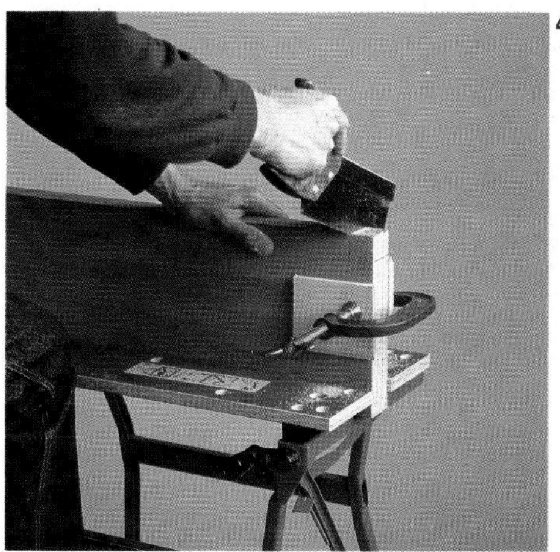

5

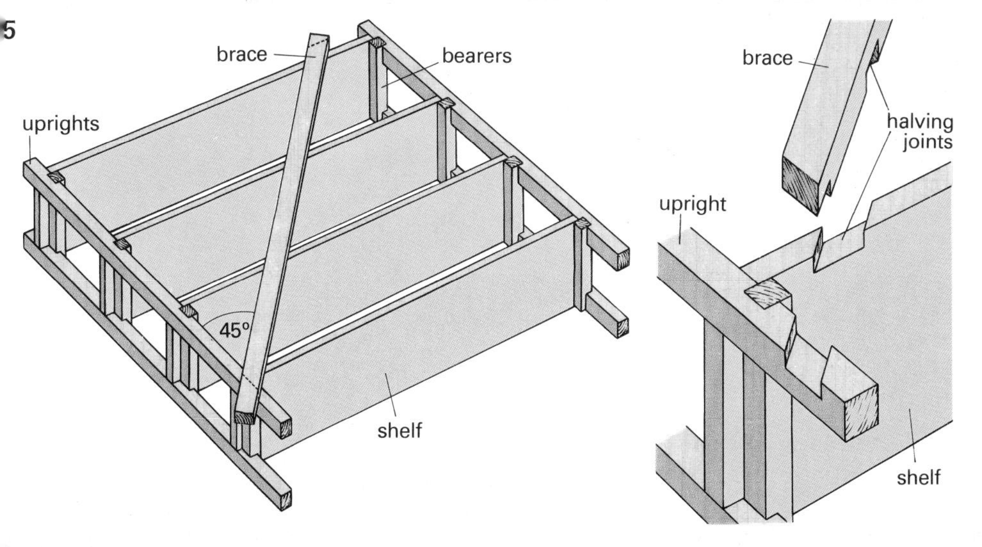

The main difference between the fixings for a normal shelf and the fixings required for a folding or drop-down shelf is that the latter needs a horizontal board on which to hinge the shelf.

When fitting a simple flap support stay, the horizontal board need be wide enough to take only the wall section of the stay. This timber, which is as long as the shelf, must be securely plugged and screwed to the wall close to the point where the stay will be attached. One screw at the top of the board and one at the bottom will be sufficient for general purposes, but for heavy weights double the number.

The shelf is hinged to a narrow timber of the same thickness as the shelf, which is screwed firmly to the top edge of the horizontal board attached to the wall. The width of this board depends on the projection of the flap stay when in the closed or down position as obviously the shelf must fall freely in front of it.

You can hinge the shelf using ordinary butt hinges set into its edges and its bearer in the usual manner. To avoid having to make recesses, you can use piano hinges, available in lengths of up to about 2m, or mount butt hinges on the underside of the shelf and support. Alternatively, use hinged brackets.

Folding table tops and worktops can be supported on wooden brackets. Timber 50 × 25mm will make a strong bracket. The right-angle halving joint at the top is made first and is screwed together, then checked with a square to ensure its accuracy. The bracing piece is then laid across the two legs of the bracket at an angle of 45 degrees. Both timbers are then marked where they cross and the depth of the joint is marked on the edge of the timber. If possible, this is best done with a marking gauge, but if you have not got one you can make a gauge by driving a countersunk-head screw into a small piece of scrap wood until the head projects just the right

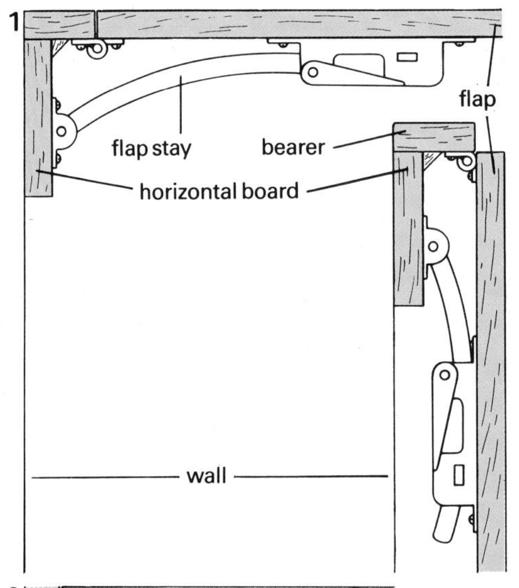

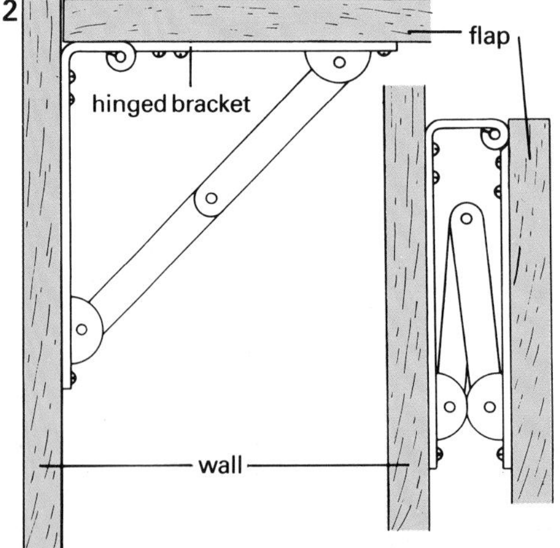

1 & 2 *Two methods of making folding shelves using flap stays or hinged brackets as supports.*

1 *Folding shelves or tables can be supported on gallows brackets.*
2 *The brackets can be made using halved and lapped joints which are glued and screwed together.*
3 *The butt hinges for the flap support brackets can be surface mounted.*
4 *Let the hinges for the flap into the bearer and the edge of the flap for a neater finish.*

amount for the depth required. Slide this gauge along the face of the wood so that the head of the screw cuts a line into the side. Then square the joint lines down to this depth line to guide you when cutting the joint.

Saw down to the depth line at each side of the joint and into the waste to make chiselling easier. Cut the waste from the face of one half of the joint and from the back of the other. Screw all the joints together and fix the bracket, with hinges, to an upright plugged and screwed to the wall.

1

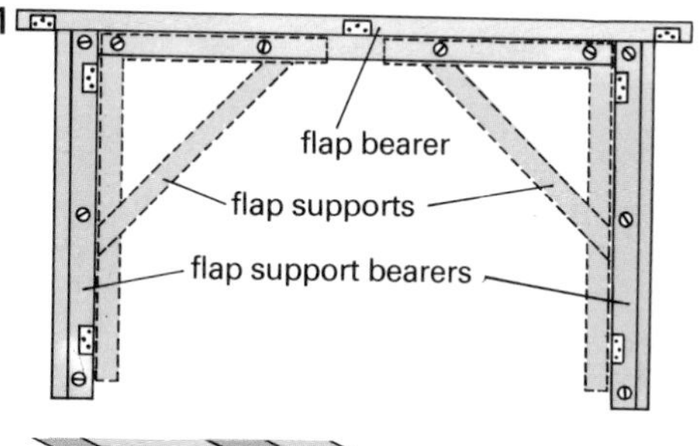

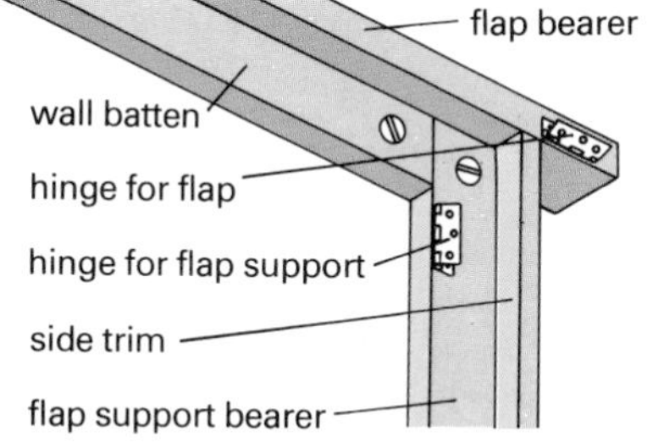

2

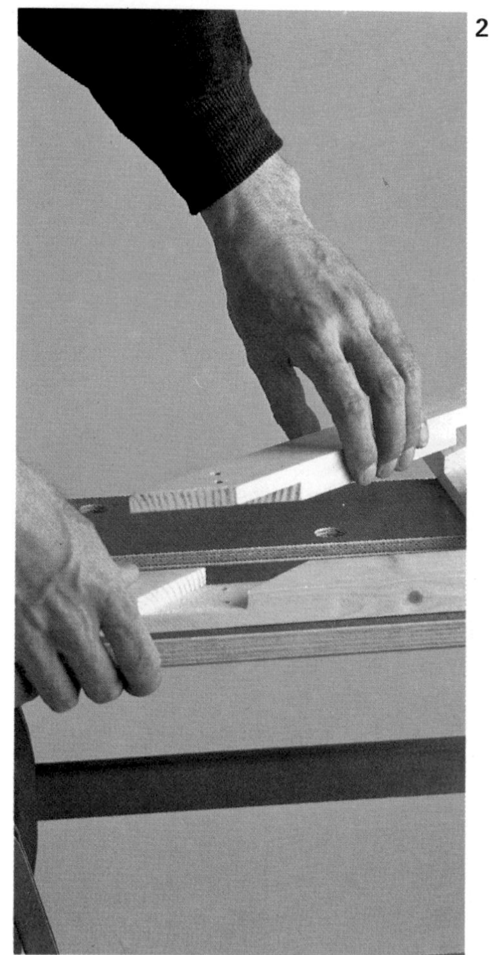

3

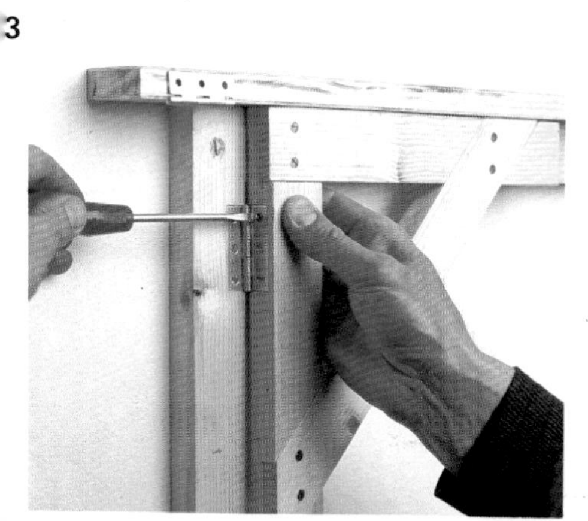

4

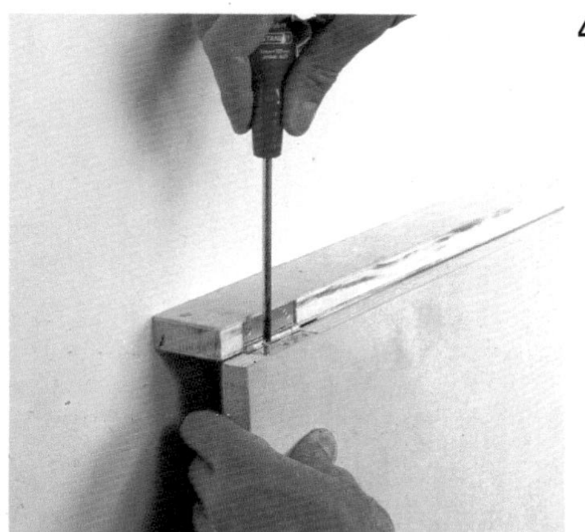

A display cabinet need be only a light wooden case with glass or wooden shelves and either with or without glass doors. As the cabinet is intended to be looked at, it is best to make it using some form of concealed joint rather than the plastic blocks which make storage cupboards easy to construct.

One of the simplest concealed jointing systems is the dowel joint. It does require some form of jig to get the holes in exactly the right places. If only a few dowels are to be used, twin-point locators are all that you need. These are metal discs with a centre point at each side. The board edge is pierced by the locator at the required position. A dowel or small tube can be used to press it home. The board to be mated is then positioned accurately and pressed on to the exposed point of the locator. Both board and locator are then separated and the pin marks are used to centre the drill. For accuracy, the holes should be bored using a purpose made dowel bit which is similar to a metal twist drill except that it has a centre point and side spurs for clean cutting and accurate centring.

When a lot of dowel joints are to be made, it is better to buy a dowelling jig. There are two or three different types but the basic principle is the same. A metal block contains holes of various sizes and set at right angles to each other. These holes are used to guide the drill for face and edge boring. Usually they take bits of 6, 8 and 10mm diameter.

Methods of adjustment for the positioning of the holes for the dowels and the sizes of timber that can be accommodated vary according to the make of the jig.

Glass shelves for ornamental displays can be supported by studs let into the side of the cabinet. Decide on the number of shelves and the adjustment needed, then drill a plywood strip with holes at the required spacings and use this as a templet when drilling the holes in the cabinet sides.

If possible, drill these holes before assembling the unit. A bush is then pressed into the hole and into this is pushed the stud on which the shelf rests. Always use a depth stop of some kind when drilling these holes, to prevent you boring too deep and damaging the face side of the boards.

A well made dowel joint will make a fairly rigid cabinet, but rigidity and appearance are improved if a back is fitted. This can be hardboard painted to match the woodwork or for better finishes use veneered plywood. If you have the facilities a rebate can be made to take the back, otherwise bevelling the edge of the board will make it less obtrusive when viewed from the side. Do make sure that the fit is very close.

1 *Two methods of fitting the back of a cabinet.*
2 *Using a dowelling jig when drilling dowel holes in the edge of a board. Note the depth stop on the drill bit.*
3 *Drilling holes for shelf support studs through a pre-drilled templet.*
4 *An exploded view of a simple display cabinet showing shelf supports and dowel joints with, inset, a detail of the pivot hinge for the glass door which needs no holes or cut-outs.*

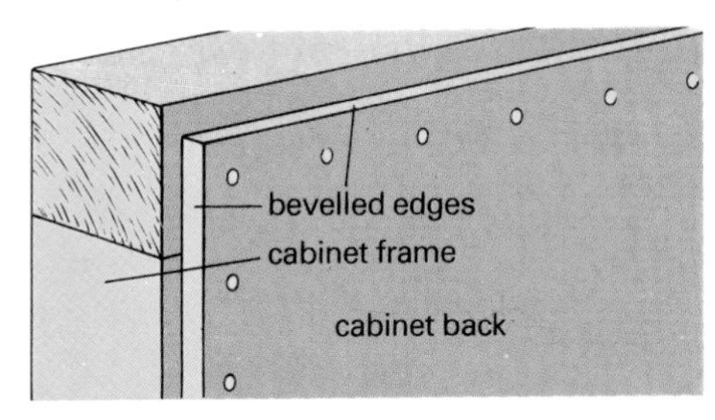

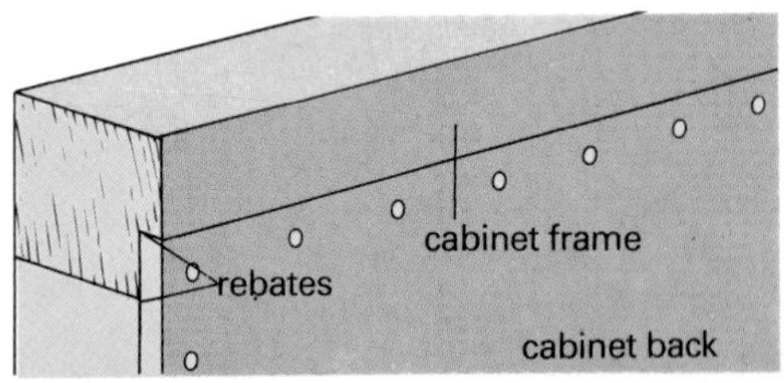

2

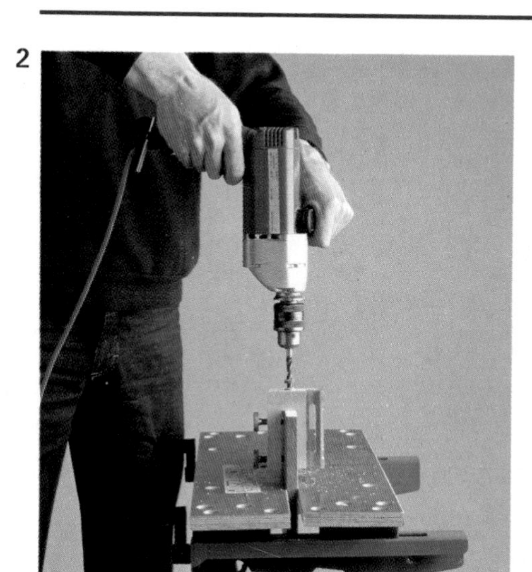

3

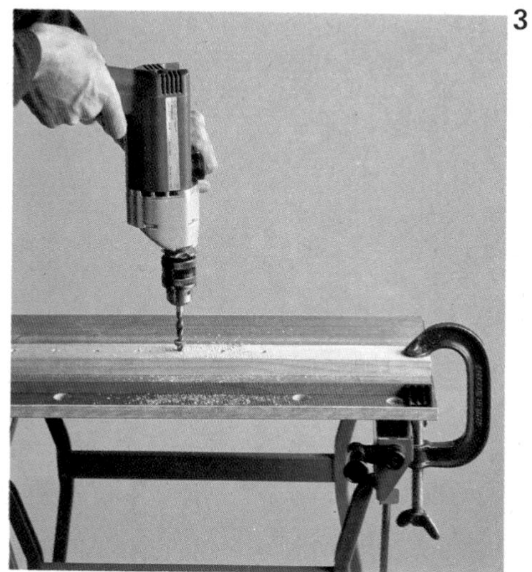

4

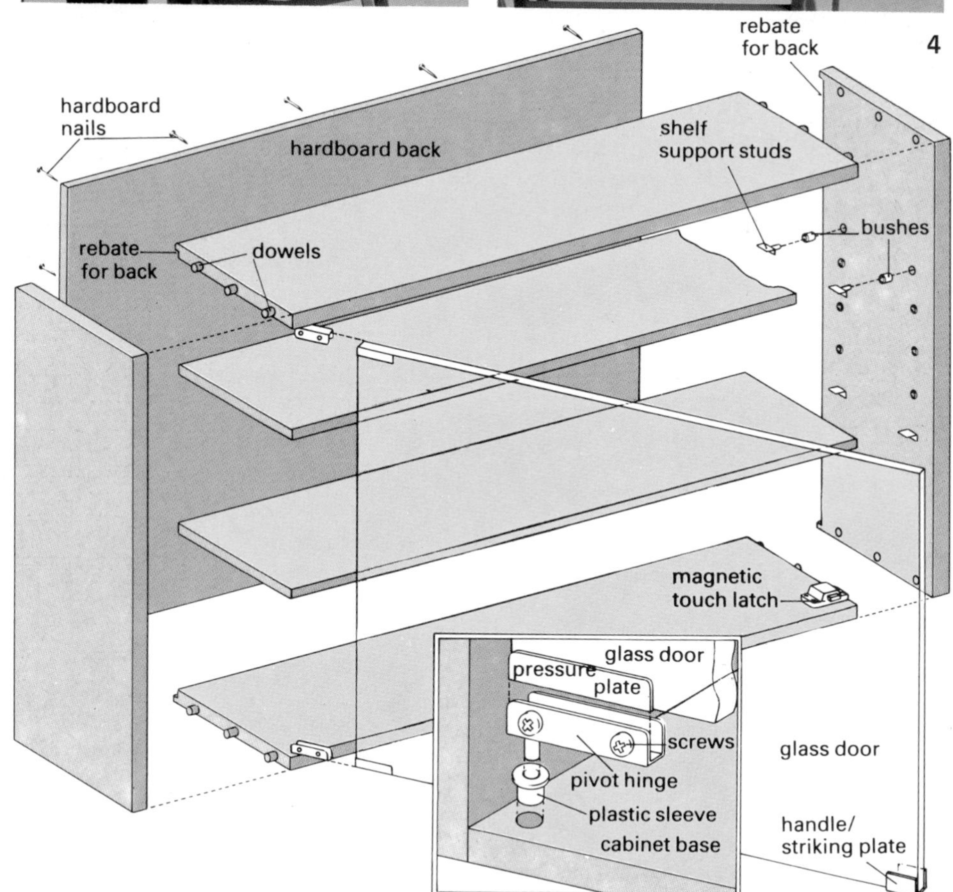
rebate
for back
hardboard
nails
hardboard back
shelf
support studs
bushes
rebate
for back
dowels
magnetic
touch latch
glass door
pressure
plate
screws
pivot hinge
plastic sleeve
cabinet base
glass door
handle/
striking plate

There are two main types of storage cabinet. One is of the quality suitable for use in living rooms and the other is the more sturdy store room type.

Cabinets for use as furniture are made using veneered boards and for these the dowel joints described in making display cabinets are most suitable. Plastic block connectors are quite suitable for general storage cabinets and for kitchen cabinets.

As these units have a fair weight of materials to support, it is more usual to house the shelves into the side of the cabinet, unless plastic blocks are being used. For the best appearance, the housing should be stopped and not cut right through to the front of the side pieces. Mark the position and thickness of the shelves, then square the lines across the boards. Mark the depth of the housing and bore two or three shallow holes at the stopped end of the housing. Clean out these holes to make a small recess and you will find this a help when sawing each side of the housing. The waste is best cut away with a chisel and finished off with a hand router. If a router is not available, you will have to take great care in getting the housing level and even.

The front edge of the shelf is cut back to clear the end of the stopped housing and the shelves are glued into place. These shelves are slid into the cabinet from the back after the top and bottom have been dowelled and fixed.

When the shelves are installed the back can be fitted. This is either set in a rebate, or if that cannot be made the cabinet must be squared by measuring the diagonals, which must be the same length, and the back nailed to the sides and bottom edges. Extra nails into the shelves will make an even more rigid job.

Rigidity is very important if a door is to be inset as any movement of the cabinet would cause the door to bind and jam. Face fitted doors hung on adjustable concealed hinges are the best for this type of unit as any slight discrepancies can be overcome by the adjustment of the hinge.

Wall-hung cabinets must be firmly fixed and this means making provision for fixing when making the cabinet as well as

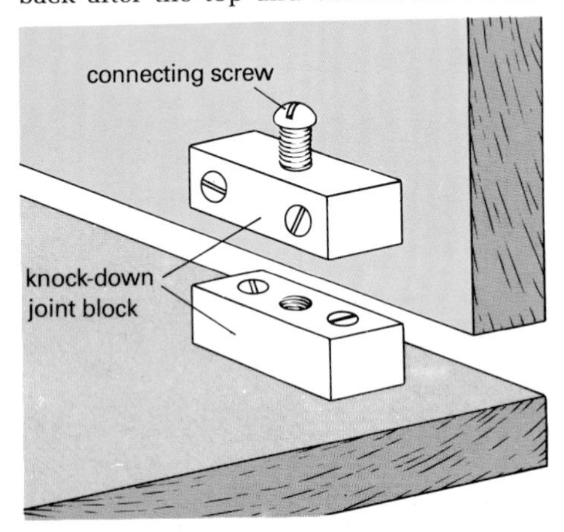

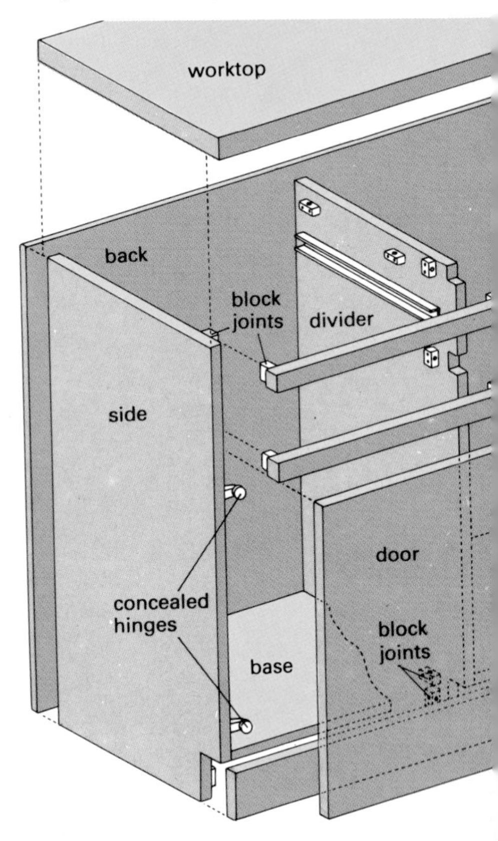

providing a secure fixing in the wall. One method for use where heavy weights are expected, is to fit a supporting batten underneath the top and the shelves. The shelves and the top can be screwed down into these battens and the battens then plugged and screwed to the wall.

Brass mirror plates, which are screwed to the edges of the top or sides and then fixed to the wall are really satisfactory only when solid timber is being used. Chipboard soon gives way at the edges.

Measurements have not been included in the illustrations as they are intended to show the methods used when constructing the complete unit.

A typical storage cabinet with details of the joints used. Although the entire unit can be constructed using plastic blocks, secret fixings such as dowel joints and housings look more professional for living room furniture. The cabinet top can either overhang the sides or be inset between them. Dowel joints can be used in either case. The back will hold the frame rigid so the shelves can be fixed in housings or they can be made adjustable by using one of the support systems suitable for bookcases.

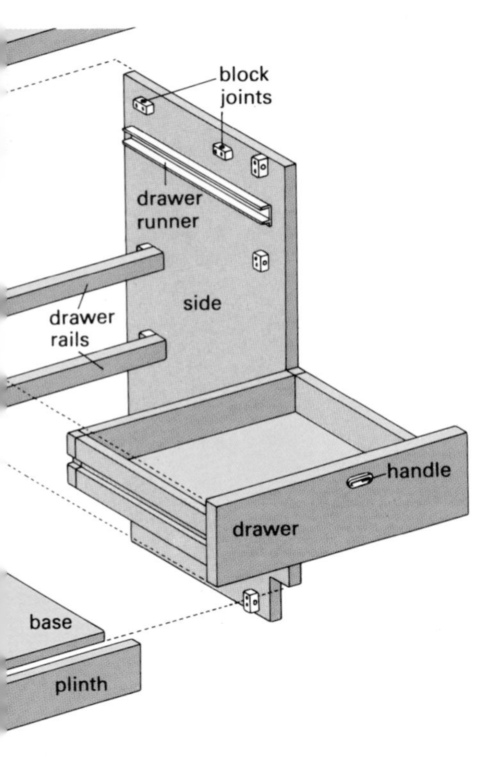

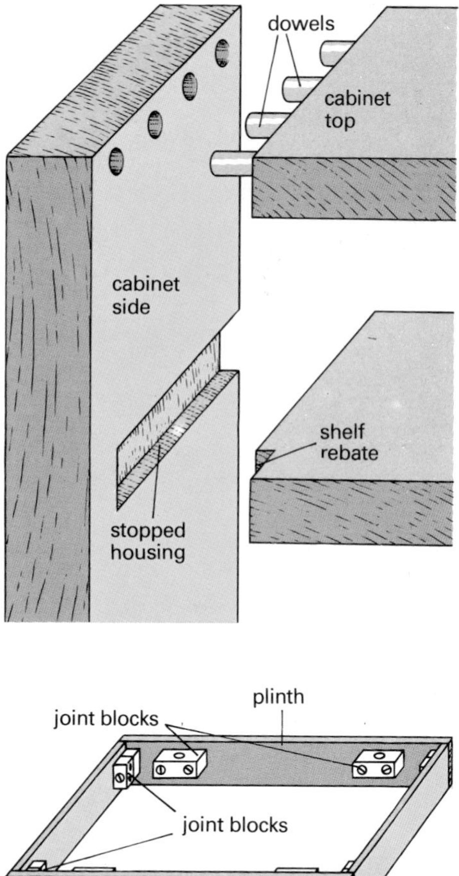

1 *The bottom of a drawer slides into grooves in the sides and then the back is fitted and fixed in place.*
2 *The bottom holds the drawer square so it must fit neatly into the back. Remember to allow for the depth of the grooves when measuring up for the size of the base.*

Making drawers can be a tedious and time-consuming job and it is unlikely that you would want to go to all the trouble of making dovetail joints for the fronts. Housed joints will do the job just as well and, if you prefer it, you can use plastic extrusions instead of wood.

There are one or two different plastic drawer systems, but most include a corner joint which enables the drawers to be assembled using only square cut materials. Using these drawer systems does limit the depth of the drawers to that of the extrusions. If you want to have a more varied choice you will have to use wood and chipboard.

The easiest type of drawer is the one with the front wider than the sides so that it fits over the face of the cabinet. The drawer itself is then made as a square box with the ends housed or rebated into the sides. These joints are glued and pinned.

Ideally, the bottom of the drawer is fitted into a groove in the sides and front, and the back of the drawer is made shallower than the sides, finishing level with the top edge of the groove so that the bottom can be slid into place and pinned to the bottom edge of the drawer back.

A rebate can be made for the drawer bottom, but it is not very good as it does not leave much for the drawer to run on. The drawer will not run freely if the bottom is simply fixed directly to the bottom edges of the sides.

Side runners are therefore the best solution. A strip of hardwood, glued and screwed to the outside of each side of the drawer, slides on a similar strip of hardwood fixed to the inside of the cabinet. This means that the drawer tray is made 20mm narrower than the opening to make a space for the two 10mm runners. The false front, which overlaps the sides of the cabinet, is screwed into place through the front of the drawer.

Using side runners means that there need be no cabinet rail showing between the drawers which will then present a flush finish. If rails are needed to

1

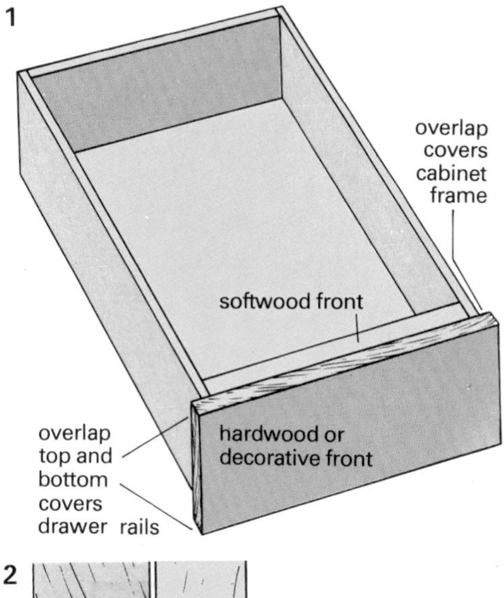

3

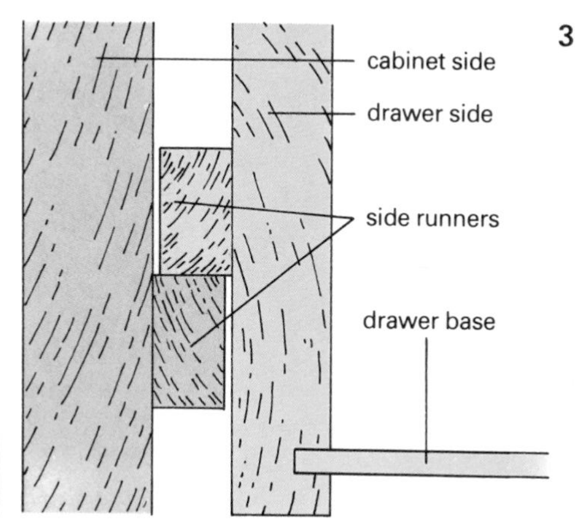

2

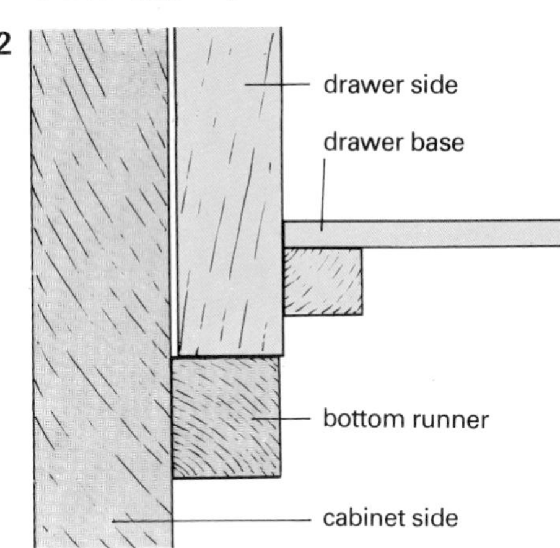

Methods of making drawers.
1 *The drawer can have a decorative front screwed to a softwood backing.*
2 *The drawer slides on a hardwood runner screwed to the side of the unit.*
3 *Hardwood runners are screwed to both the sides of the unit and to the sides of the drawer.*
The former method is often used where front rails form a feature of the unit. In the latter method the two hardwood runners wear better and the drawer fronts are not separated by the front rails of the cabinet.
4 *This illustrates how drawers are made-up using one of the plastic drawer making systems which are generally obtainable as complete kits.*

strengthen the cabinet, they can be fixed by letting them into the sides or screwing them through the sides, dependent on the quality of finish required. In this case the width of the drawer front will have to be adjusted to cover the rails.

Drawers can be run on bottom rails if the drawer bottom is fixed to battens glued and pinned to the inside of the drawer about 10mm above the bottom of the sides, but this reduces the storage depth of the drawer.

Always remember to make your drawers as strong as possible so that the weight they bear will be well supported.

4

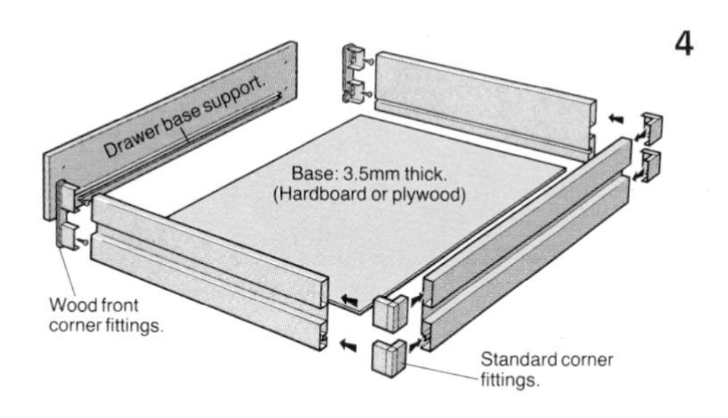

Making a corner unit presents few problems as the sides can be screwed to each other at the back, as well as being screwed into the triangular top and bottom. When the unit is in position these screws will not show. The shelves can also be screwed into place through the sides; there is no need to house them if they will have little weight to carry. If preferred, they can be supported on studs so that they can be adjusted whenever necessary.

Corner display shelves are usually open at the front, but if a door is required a 50 × 25mm frame can be made and secured to the unit by plastic blocks behind the top and bottom rails. The doors can then be hung on butt hinges or piano hinges, or for glass doors pivots can be used.

As the sides of the cabinet will be of plywood or chipboard, they can be screwed directly to the wall. Drill a hole near the top at each side, then hold the unit in place and mark the wall. Drill the wall and fit plugs, then screw the unit into place.

Floor standing kitchen units are quite large as the units which they adjoin are about 600mm deep. This large corner area where two runs of floor cabinets meet, does present a problem of access. The easiest solution is to provide the corner unit with a lift up lid. The unit itself is made as a simple box joined at the corners by plastic blocks. The bottom, which can

A corner cabinet is constructed in the same way as an ordinary cabinet but the door can be a bit more awkward to hang. There are also more sawn edges to finish with veneer or plastic strip.

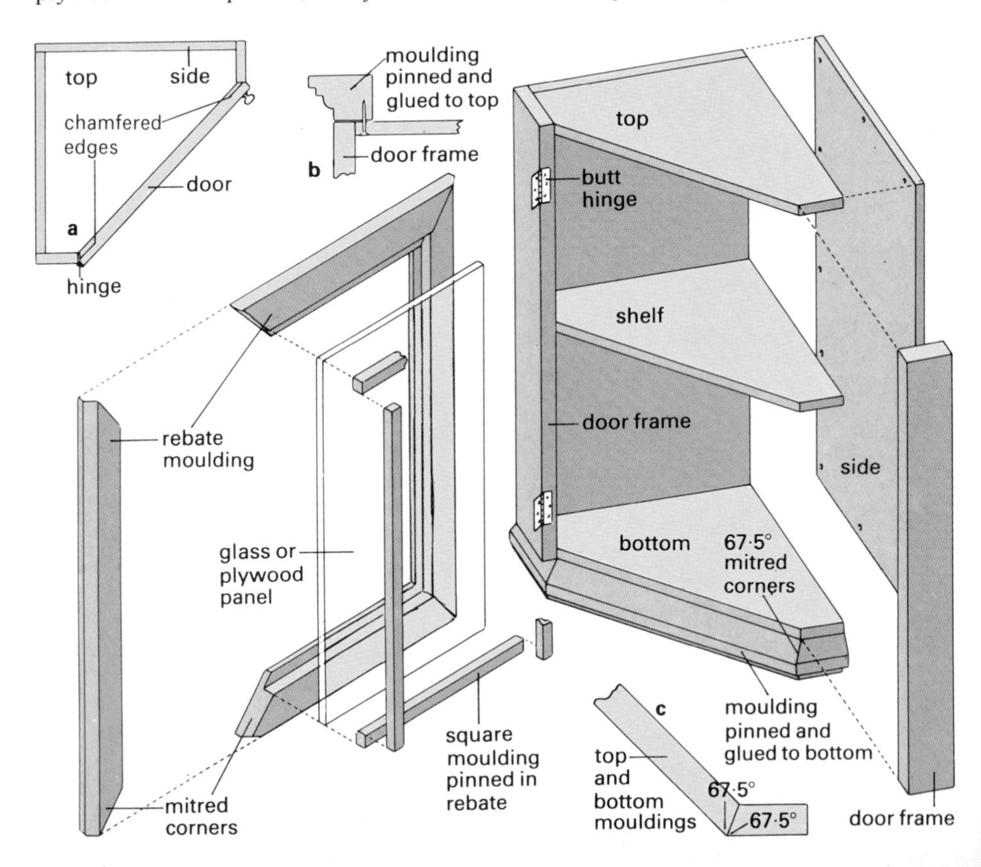

be made of chipboard, holds the unit square and can be supported on 25mm square bearers fixed to the sides, just above floor level.

A lift-up top is most easily provided by fixing a strip of worktop about 75mm wide to the top of the unit using plastic blocks. The flap can then be hinged to this strip by means of a piano hinge. The unit can then be fitted into place and screwed to the wall. It can be used for storing large items or as a linen box.

In order to get a front door access, the unit must be made a little bigger than the depth of the kitchen units. The larger it is made the bigger the door that can be provided. A scale drawing, or a full-size drawing on the floor would help in ascertaining the size of opening needed to provide access for large items.

Construction is the same as that described for other cabinets, except that the top is fixed to the sides by small metal brackets. The doors can be hung one at each side so that they close to make an internal right-angle, or they can be hung one onto the other, like folding doors. In each case they are held closed by magnetic catches.

Shelves or a carousel can be fitted in the usual way. A carousel is a set of circular shelves on a central pivot which enables you to revolve them to get at the contents easily.

Corner units in a run of kitchen units can be a bit of a problem. Here are two ideas for getting the most out of the space available. The lidded type fits the corner exactly. The door type is bigger.

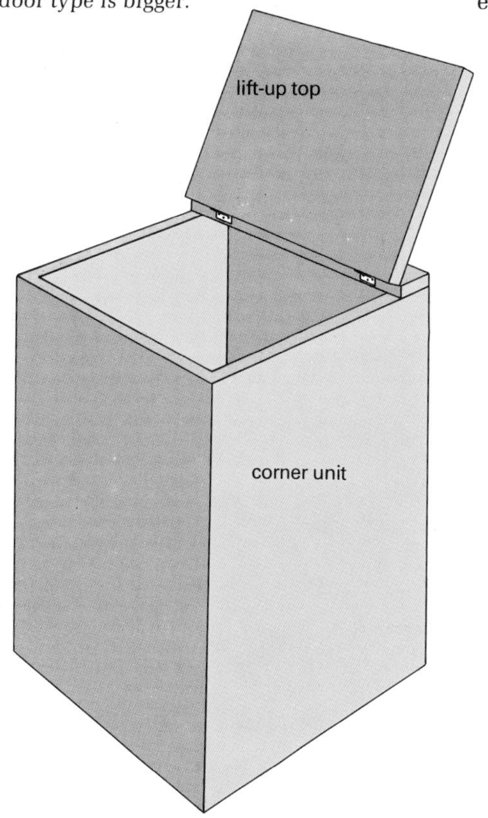

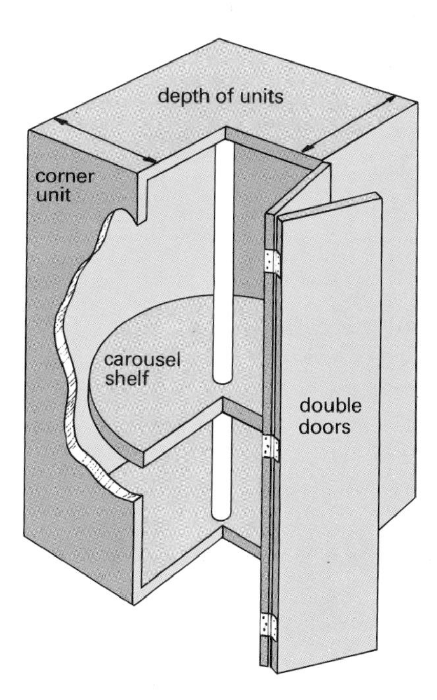

All articles made of wood need treating with a preservative or finish, not only to preserve and protect the surface, but also to bring out the inherent beauty of the grain and the texture of the timber. The quality of finish is extremely important, as it is by this that the work is usually judged. Although painting will hide any slight surface defects, any blemish in wood becomes even more noticeable when a clear finish, or a stain and clear finish are applied. It is important also that all woodwork, except that which is being treated with a preservative such as creosote, is perfectly clean and smooth.

The main processes in finishing are filling, stopping, staining and applying the finish. When a clear finish is to be applied, it is essential when an orbital electric sanding machine has been used that the surface is still finished afterwards by sanding with the grain, by hand. If this is not done, small circular scratches resembling fish scales will be seen in the final coat.

Preservatives

Fences and sheds can be treated with creosote to BS 3051. If the brown colour is not acceptable, preservative of another colour can be used. Where the timber will be in contact with plants, such as on the inside of greenhouses and forcing frames, a green horticultural preservative should be used. Colourless preservatives are also available, for use on interior and exterior woodwork; these can be painted when dry.

Painting

Emulsion paint can be applied direct to sanded wood and no knotting, primer or undercoat is required. Before applying solvent-based paints, any knots in the wood should be sealed with pure shellac knotting and allowed to dry. Any cracks or dents should then be filled with wood-stopping, which should be applied slightly proud of the surface to allow for shrinkage and, when dry, papered smooth. Next, a pink or white primer, followed by an undercoat of a similar colour to the top coat should be applied. If a first-class finish is

required, lightly rub this smooth with fine glasspaper when it is dry, and apply a second coat. This too can be rubbed down before applying the finishing coat.

Because paint is pigmented it does not usually have as good a flow as varnish and, to make sure that no brush marks are left on the surface, it should be 'laid-off' to a greater extent than a varnish. The paint should be applied in one direction, then with a slightly lighter pressure at right angles, then with lighter pressure still diagonally, finishing off with the minimum pressure, drawing the brush lightly across the surface in the original direction of application.

The best results are always obtained by using good quality brushes of the correct size. For example, it is no good using a 1″ brush for painting a door – a 2½″ brush would be more suitable. Immediately after use, brushes should be cleaned with white spirit or a proprietary solvent.

Exterior wood-stains can be used on timber instead of paint. These are not to be confused with the transparent wood-stains or dyes used for staining wood prior to the application of clear finishes. They contain a pigment and tend to obliterate the grain, but they are easier to apply and maintain than paint.

Left *Different finishes and veneers. Top to bottom: Gloss polyurethane on rosewood; Matt polyurethane on sen; Yacht varnish on padauk; Plastic coating on burr walnut; Teak oil on teak; Danish oil on Baltic pine; Satin polyurethane on coromandel.*

Below *Selection of dyes on sycamore. Top to bottom: Light oak; Dark Burmese teak; Light Scandinavian teak; Dark oak; Walnut; Medium oak; Brown mahogany; Ebony; Red mahogany; Pine.*

Non-pigmented finishes

All finishes alter the colour of wood to some extent and some woods, for example, mahogany and walnut, turn much darker even when a completely clear finish is applied. An approximate idea of the colour the wood will become when finished with a clear solution can be seen by damping a small area with water. If this colour is too light, then the wood can be stained before finishing. It is only possible to stain wood to a darker colour; for a lighter shade it must be bleached.

When staining wood, it is advisable to test the stain on a spare piece of wood, or on an area which would not normally be seen, as it is difficult to remove stain which has been recently applied. If the wood has an open grain, and a smooth finish is required, then a grain-filler should be used for filling the pores, or extra coats of the finish would have to be applied and then rubbed down with an abrasive paper. Any cracks or holes in the wood should be filled with wood-stopping before staining.

The final finish may be of a type which gives a surface film, such as French polish, varnish or polyurethane. The last two are available in gloss, satin and matt finishes. Varnish stains are also available which will

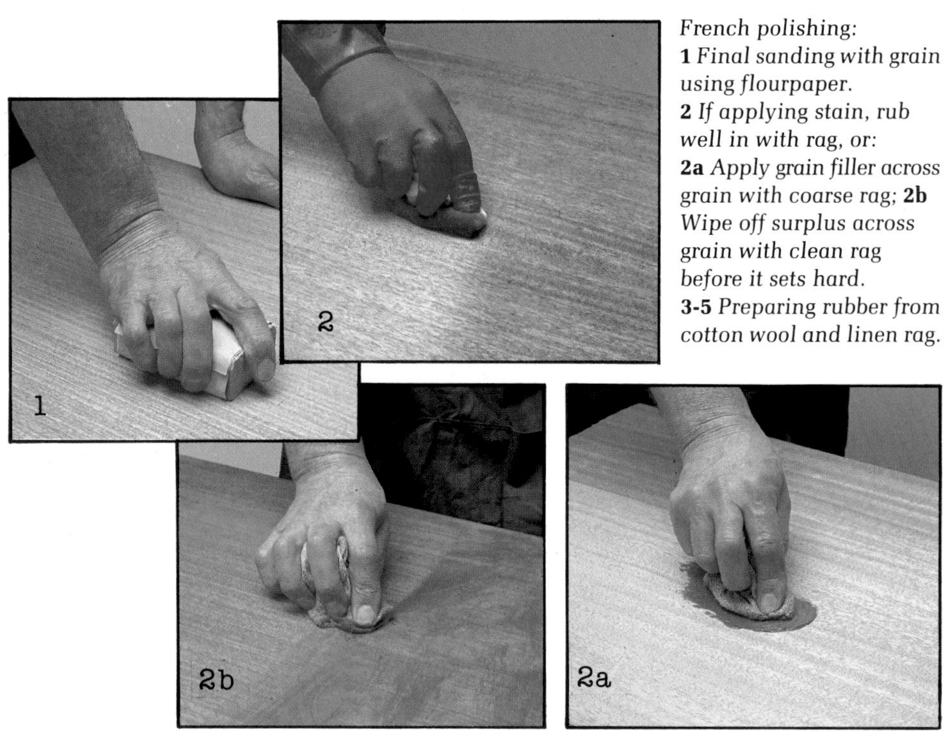

French polishing:
1 *Final sanding with grain using flourpaper.*
2 *If applying stain, rub well in with rag, or:*
2a *Apply grain filler across grain with coarse rag;* **2b** *Wipe off surplus across grain with clean rag before it sets hard.*
3-5 *Preparing rubber from cotton wool and linen rag.*

colour and finish the wood in one operation. It is important to note, however, that each extra coat of varnish stain will darken the colour and, unless brushed out very evenly, the colour will vary as the thickness of the film varies. When wood is stained with a penetrating dye, the colour will not vary however many coats of clear finish are then applied. When varnishing an external door, it is important that at least one coat is applied to the top, bottom and side edges to prevent water being absorbed at these points, which would eventually cause the varnish to fail.

Alternatively, oiled finishes may be used, such as teak oil and Danish oil. These finishes are far easier to apply than the previous types as they are merely wiped over the surface with a cloth. Teak oil leaves the wood with a soft, lustrous finish. When a high gloss finish is required on exterior woodwork, a yacht varnish should be used. These usually contain tung oil which has outstanding exterior durability.

Waxing

This type of finish is popular on wood such as pine and oak. Before applying wax the wood should be sealed with French polish if a golden colour is required, or with transparent French polish, which will not alter the colour of the wood. Two coats of French polish should be applied with a brush or rag and, when dry, lightly rubbed down with fine glasspaper. The surface can also be rubbed down with fine steel wool and wax polish. This treatment will give an acceptable satin finish. If, however, a higher gloss is required, then the wax polish should be applied with a soft cloth or a shoe polishing brush and allowed to harden. The surface should then be buffed with a soft yellow duster or soft shoebrush.

French Polishing

For further information on this process send a stamped addressed envelope to the manufacturer of the product.

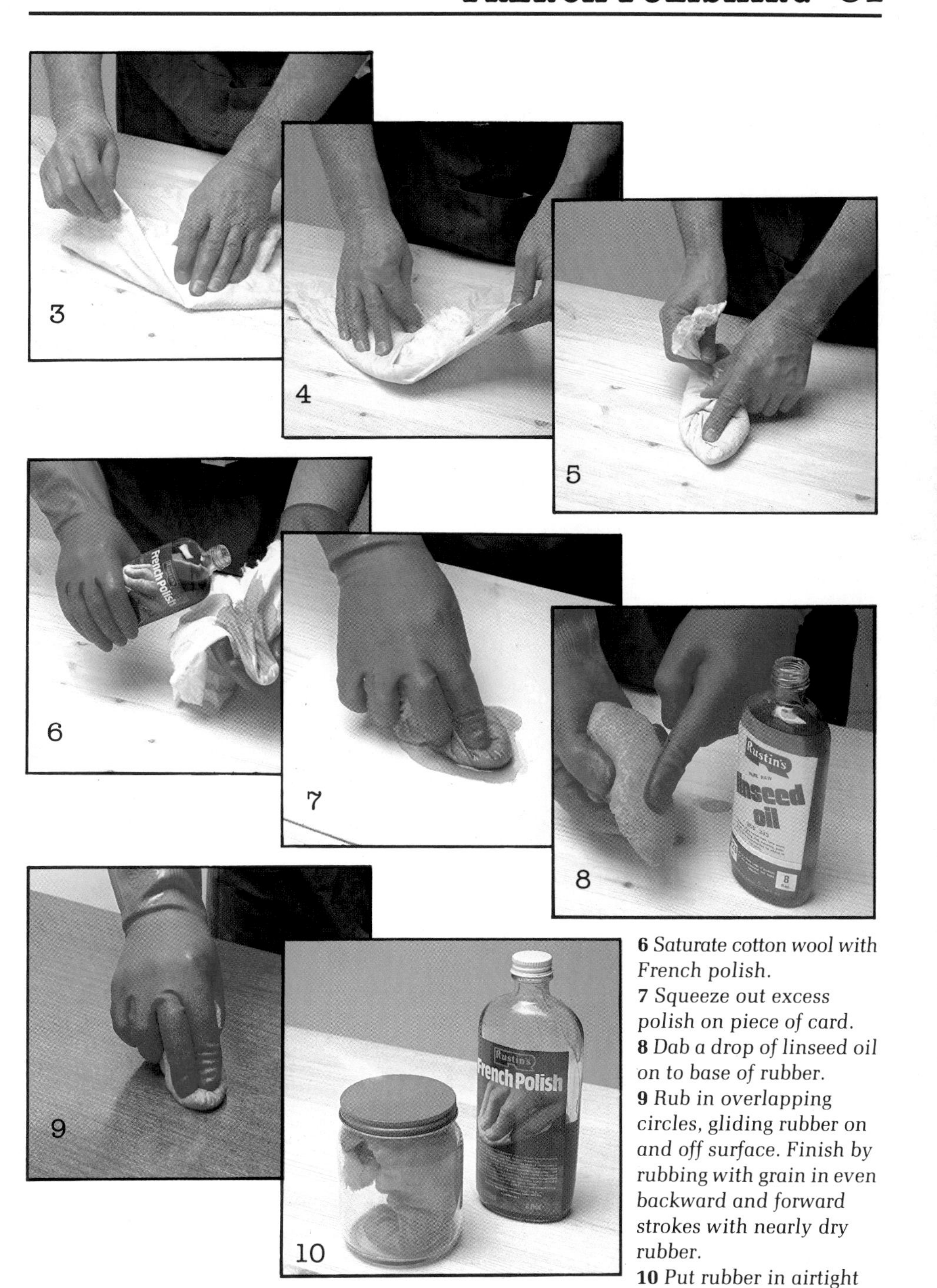

6 *Saturate cotton wool with French polish.*
7 *Squeeze out excess polish on piece of card.*
8 *Dab a drop of linseed oil on to base of rubber.*
9 *Rub in overlapping circles, gliding rubber on and off surface. Finish by rubbing with grain in even backward and forward strokes with nearly dry rubber.*
10 *Put rubber in airtight jar between applications.*

These bookshelves are a free-standing floor unit with some adjustable shelves. Construction of the shelves is in pine-board, which can support greater loads than the same thickness of chipboard.

The overall width of the unit is 900mm and its height is 1200mm. This means that two standard 1200mm boards will make the sides, while the shelves and top can be made from standard 900mm boards. The top is 36mm longer than the shelves as it goes over the top of the sides. The top and bottom are fixed by means of one-piece joint-blocks. One middle shelf is also permanently fixed in the same way, but the other shelves are adjustable. Alternatively, the unit could be assembled using dowels, with the fixed shelf set in housing joints in the sides. These techniques are described earlier in the book under Dowel & Halving Joints and Housing, Butt and Mitre Joints.

One-piece joint-blocks are neat, plastic, triangular blocks which are fixed by one screw into each panel. A plastic lid then covers the screws to provide a good finish to a sound and permanent joint.

The depth of the bookshelves depends on the type of books you intend to store on them. Paperbacks need shelves 150mm deep; the larger hardbacks require shelves about 250mm deep. If the sides, and top and bottom shelves, are made wider than the middle fixed and adjustable shelves, you can fit grooved sliding-door track to take glass or perspex doors.

The unit has a back made from hardboard and this strengthens the framework and makes it more rigid. This is necessary because it is free-standing. Having cut the sides to the required lengths, place them with their inside faces together and mark the position of the bottom and the fixed middle shelf on the edges. Then square the marks across the boards and screw the joint-blocks into place ready for the shelves.

Next, carefully cut the bottom and middle shelves to the same length and screw them to the joint-blocks. The top, which is long enough to cover the tops of the sides, is

Cutting list (millimetres)

Pine-board:	
Top	900 × 225
Sides (2)	1200 × 225
Bottom shelf, Middle fixed shelf	864 × 225
Adjustable shelves (2)	854 × 225
Plinth	864 × 75
Hardboard:	
Back	1122 × 894

Other materials:
16 Joint blocks (rigid) and screws
4 Bookcase strips, 914 long, and 8 supports
¾″ Hardboard pins
Polyurethane (satin finish)

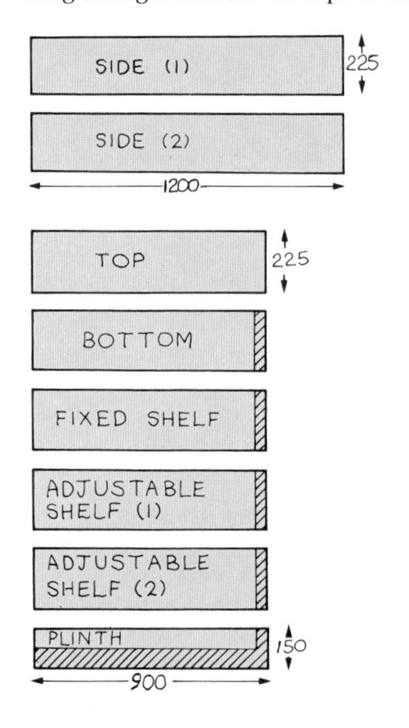

SHAKESPEARE
MY SECRET LIFE WALTER
LESLIE THOMAS THE MAGIC ARMY
MARATHON MAN
LISA ALTHER KINFLICKS
SHARP PRACTICE John Farris
ONE CHILD Torey L. Hayden
THE GARDENING YEAR
HORSE & PONY
STRANGE BUT TRUE
GERMAN SUITE
REPAIRS INSIDE THE HOME
THE UNIVERSAL STORY
THE MGM STORY
PORSCHE
FERRARI
MODERN GUNS
NEW YORK
CYD CHARISSE
VOGUE
ANTIQUES
THE COMPLETE BOOK OF BABYCARE
ROAD ATLAS of Great Britain
CUISINER'S MENUS
GARDENS OF THE NATIONAL TRUST
THE RAILROADERS
WINE
GEORGE RAINBIRD

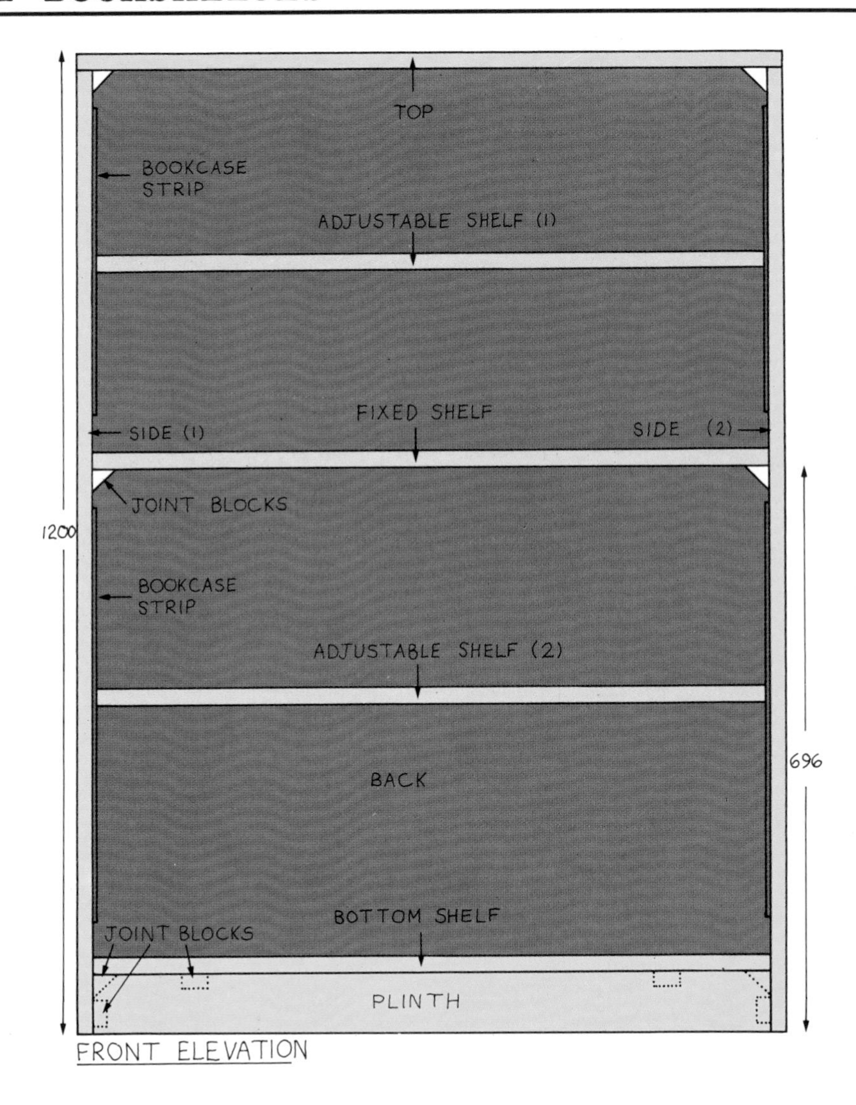
TOP
BOOKCASE STRIP
ADJUSTABLE SHELF (1)
FIXED SHELF
SIDE (1)
SIDE (2)
JOINT BLOCKS
1200
BOOKCASE STRIP
ADJUSTABLE SHELF (2)
BACK
696
BOTTOM SHELF
JOINT BLOCKS
PLINTH
FRONT ELEVATION

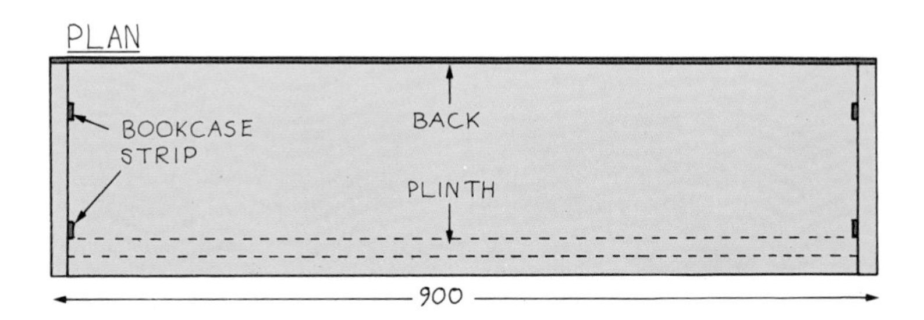
PLAN
BACK
BOOKCASE STRIP
PLINTH
900

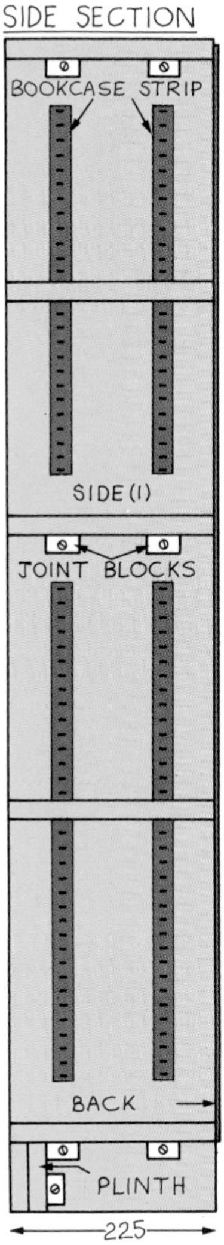

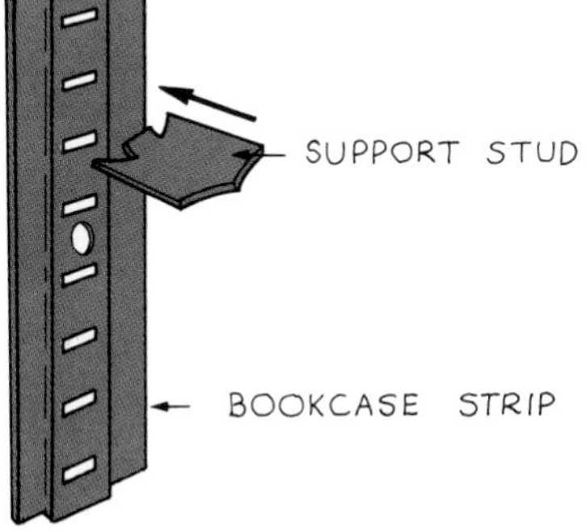

fixed in the same way.

With the framework constructed, the metal bookcase-strips for the adjustable shelves can be screwed into place 50mm in from the front and back edges. They are designed to fit on the surface of the wood and do not need a rebate. When cutting the adjustable shelves to length you must allow clearance for these metal strips at each end of the shelves. The shelf-support clips engage in slots in the metal strip so it is important that these slots are in line at each side and at the back and front of the unit.

If you cut the hardboard squarely you can use it to square up the whole structure. It is held in place by 20mm hardboard pins. If you are using a clear finish such as polyurethane or Danish oil, then the inside of the hardboard back should be sealed and painted in a colour to suit the board finish before it is nailed into place. If the unit is to be painted then the hardboard can be sealed before it is fixed and then painted with the rest of the unit on completion.

At the bottom of the unit there is a plinth or kicking-board set back about 25mm to make it less vulnerable to scratching. This can be made from the same type of board as the rest of the unit, but as it is only 75mm wide you might find that stock sizes are a little wasteful, so it can be made from a length of matching timber.

To avoid damage to the plinth you can treat the timber with a plastic coating or face it with a laminate, either in a matching wood grain or a contrasting plain colour. Black makes a good plinth colour and it goes with most types of wood. The plinth is held in place by screwing one joint-block to each end and a couple under the bottom shelf, evenly spaced along its length.

Working with pine-board

Because pine-board is made from strips of timber glued together, it may swell and shrink with changes in atmospheric conditions. Where possible, therefore, it is advisable to apply one coat of the chosen finish to seal the timber before assembly.

However much storage space you already have, there always seems to be a need for more shelves. The construction of this small unit demonstrates a very simple technique for making modular shelving units. They can be of almost any length, width or height – the techniques for making larger units being identical. The unit shown is made with pine uprights and Contiplas shelves. Contiplas should have a maximum span of 700mm between supports for any loadbearing shelves. This 700mm can be increased if the shelves are strengthened by battens at the front or back.

All pieces of the same length and all holes must be marked out in a group for maximum accuracy. After marking out, saw the legs and cross-rails to length and bore all holes for the Scan fittings. The technique for using Scan fittings is explained in the instructions for making the coffee-table (see pages 62-65). Then clean up the legs and cross-rails and assemble using wood-working adhesive. Glasspaper well and finish with polyurethane.

Now cut the three shelves to length. The top shelf is fixed to the three frames using rigid joint blocks. The two lower shelves are screwed through into the frames and screw caps set into counter-bored holes are used to cover the heads.

Finally, cut the two backs to size and screw them into position between the frames using 1¼″ No. 6 screws.

Cutting list (millimetres)

Contiplas:	
Backs (2)	600 × 380
Shelves (3)	1200 × 230
Pine:	
Legs (6)	755 × 47 × 22
Cross rails (9)	230 × 47 × 22

Other materials:
18 Scan bolts, nuts, and cover-heads
6 Joint blocks (rigid) and screws
$1\frac{1}{4}''$ × No. 6 Chipboard screws
Plastic screw caps
Woodworking adhesive
Polyurethane

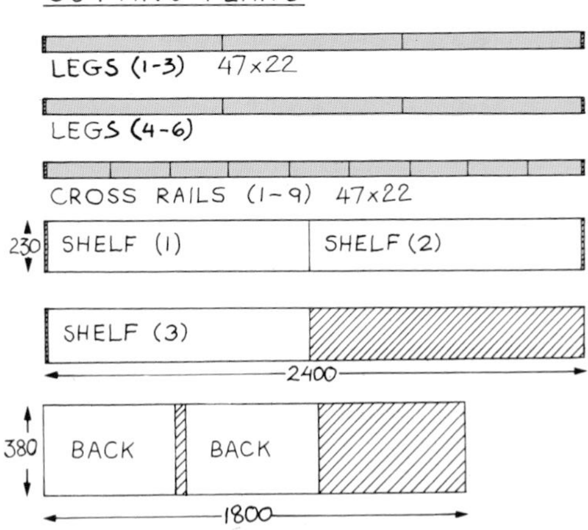

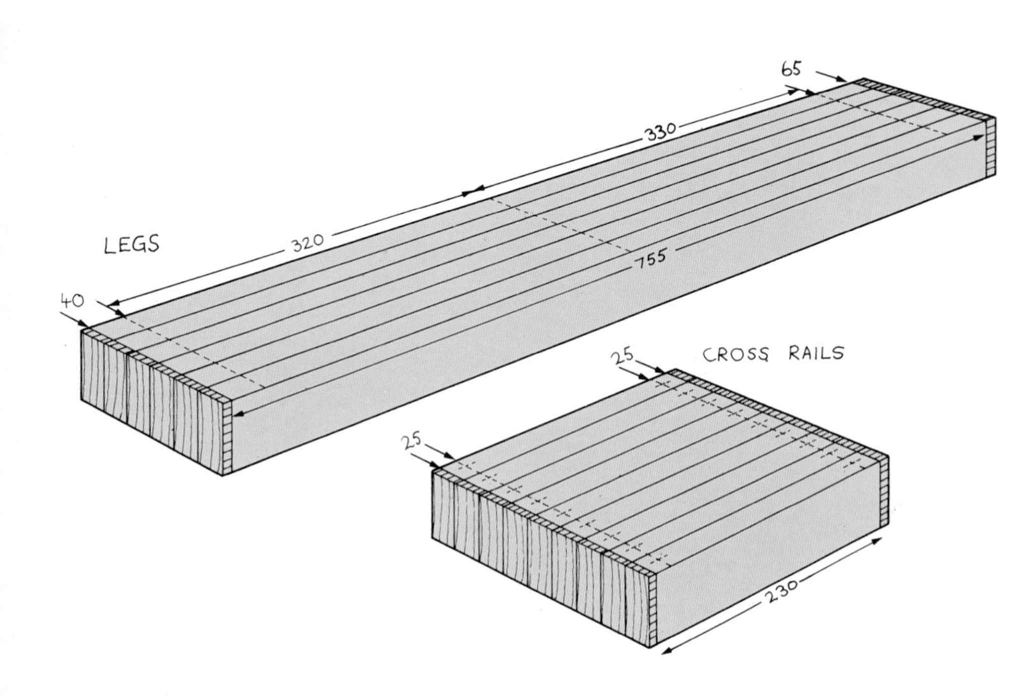

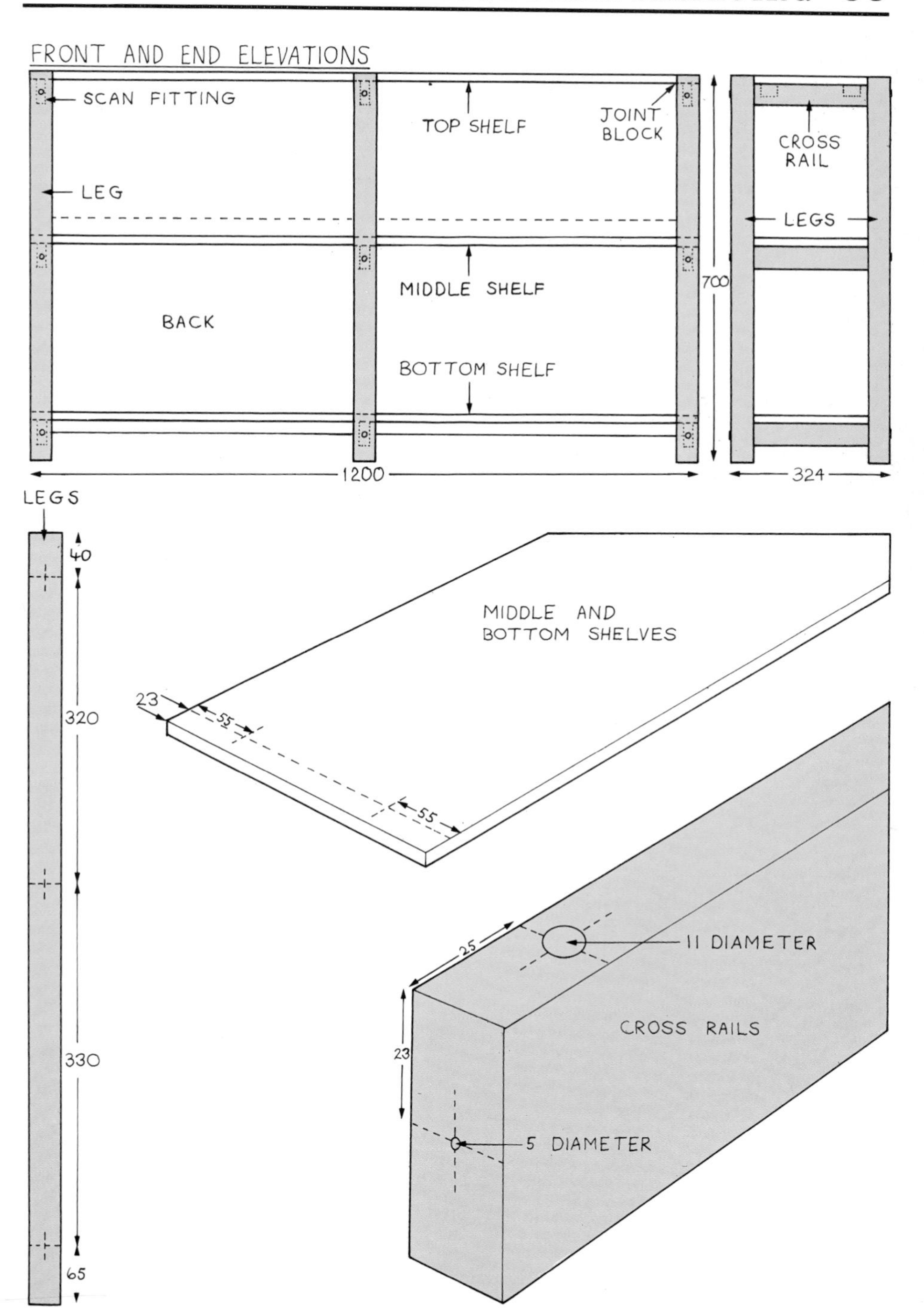
FRONT AND END ELEVATIONS
SCAN FITTING
TOP SHELF
JOINT BLOCK
CROSS RAIL
LEG
LEGS
MIDDLE SHELF
BACK
BOTTOM SHELF
700
1200
324
LEGS
40
320
330
65
MIDDLE AND BOTTOM SHELVES
23
55
55
25
11 DIAMETER
CROSS RAILS
23
5 DIAMETER

BATHROOM CABINET

This bathroom cabinet is made from Contiplas which is suitable for the damp atmosphere of a bathroom and is easily wiped clean. There is room to store most of your bathroom requirements and it has a useful-size mirror attached to the outside of the door. There is also a hinged flap for putting items on while you are using them.

First, cut the Contiplas panel into the sizes shown in the cutting diagram, allowing an extra 20mm on the length of each. Then cut the 600 and 273mm long panels in half lengthways to form the shelves and sides

Place the two sides together in a vice or cramp, and mark them to length, with a marking knife. Square this cut line right round the panels, leaving approximately 10mm spare at each end. Repeat this operation to mark the top, bottom and two shelves to length, again leaving 10mm spare at each end.

Once the cut lines have been squared round the ends, all the panels can be cut to size. For this job use a panel saw with approximately 9 teeth to 25mm. Clean up each end using a sharp plane and then edge veneer the two side panels at both ends. There is no need to edge the shelves.

Next, mark the position of the screw holes in the sides of the cabinet. Bore these and then counter-bore to a depth of 4mm and 10mm diameter for the screw caps.

The cabinet can now be assembled. Cut the back to size and paint it to the colour required using emulsion paint, but do not fit it yet. Cut the door and flap to size and veneer the exposed edges. The door and the flap can now be hinged and the magnetic catches fitted, followed by the flap stays and the shelf studs.

Now fit the back using $\frac{3}{4}''$ countersunk screws, driving them well in. Attach the handles and mirror, and screw the cabinet to the wall using two mirror plates. These are screwed into the back top edge of the cabinet with countersunk chipboard screws. The third hole in the plate is used to attach the cabinet to the wall using screws at least $1\frac{1}{2}''$ long.

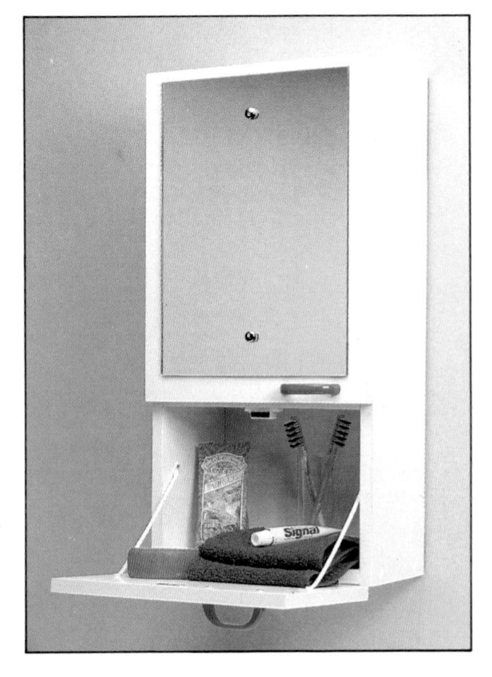

Cutting list (millimetres)

Contiplas:	
Sides (2)	600 × 150
Top	273 × 150
Bottom	273 × 150
Shelf, fixed	273 × 150
Shelf, adjustable	273 × 150
Door	393 × 303
Flap	205 × 303
Hardboard:	
Back	600 × 303

Other materials:
2 Handles
4 Hinges, single-cranked, and screws
2 Flap stays, plastic
2 Magnetic catches
4 Shelf studs
1 Mirror, 350 × 255, and mirror screws
$1\frac{1}{2}''$ × No. 6 Chipboard screws
$\frac{3}{4}''$ × No. 4 Countersunk woodscrews
Plastic screw caps
Edging strip
Emulsion paint (for back)
Wall-fixing screws and plugs
1 Cupboard door lock (optional)

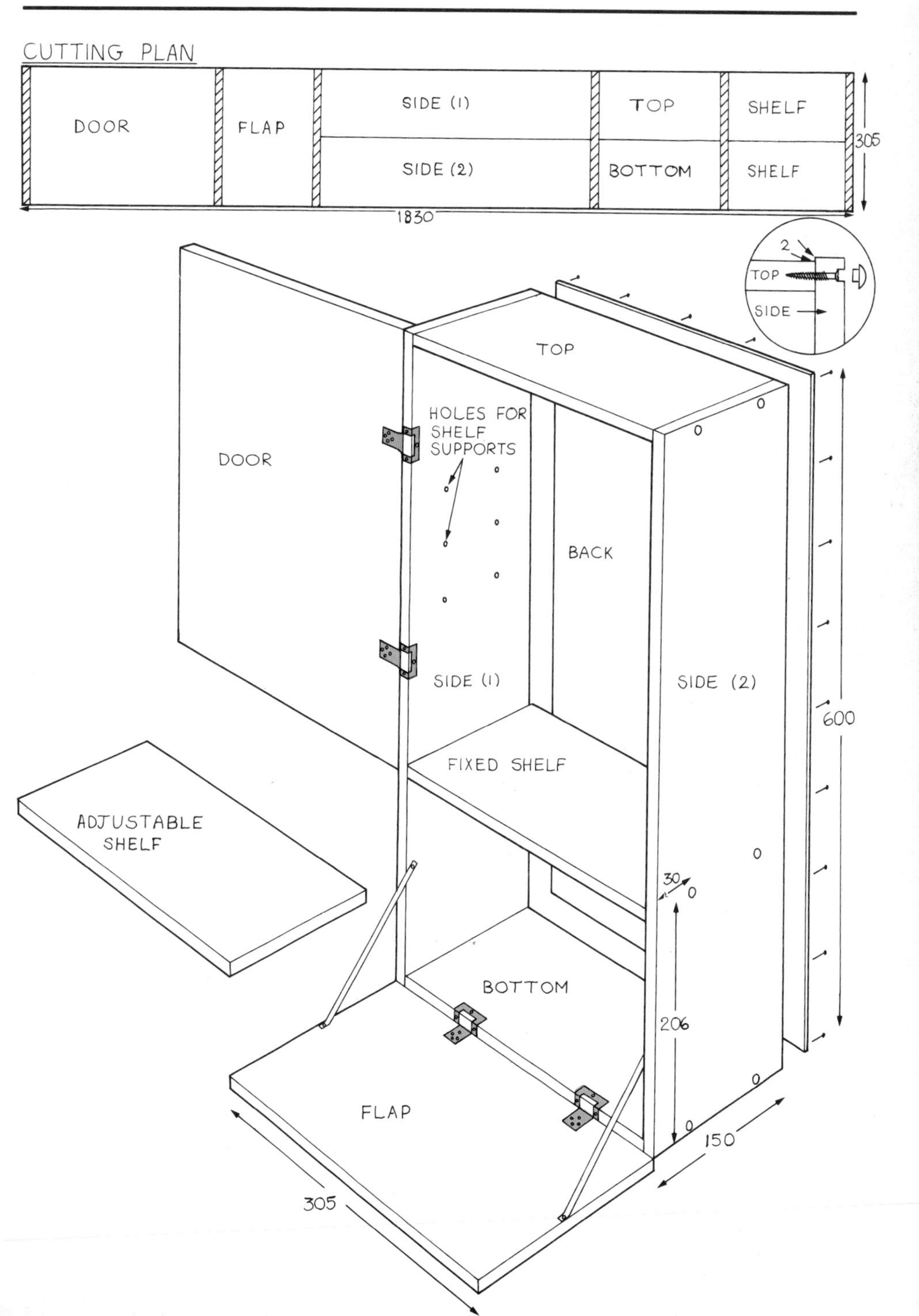
CUTTING PLAN
DOOR
FLAP
SIDE (1)
SIDE (2)
TOP
BOTTOM
SHELF
SHELF
305
1830
2
TOP
SIDE
TOP
HOLES FOR SHELF SUPPORTS
DOOR
BACK
SIDE (1)
SIDE (2)
600
FIXED SHELF
ADJUSTABLE SHELF
30
BOTTOM
206
FLAP
150
305

COFFEE TABLE

This low table with a shelf for magazines is a useful piece of furniture for any living-room. The legs are made from timber and the top and shelf from veneered chipboard.

To ensure that each set of component parts matches, the marking out should always be done by the 'group' method as follows. Hold the four legs together as a group in a vice or cramp and mark these out as shown in diagram A. Cut lines, made with a sharp knife, are always used when a saw cut is to be made; all other lines are made with a pencil, sharpened to a chisel point. On the centre of each pencil line mark the position for boring the hole for the Scan bolt (5mm diameter), and bore these holes. Then saw off the waste.

Next, place the four long rails together and mark these as shown in diagram B. The four long rails are identical except for the position of the Scan bolt which goes through the thickness of the rails. On the

Cutting list (millimetres)

Contiboard:	
Top	910 × 380
Shelf	700 × 286
Timber (to choice):	
Legs (4)	360 × 32 × 32
Long rails (4)	700 × 45 × 20
Short rails (4)	286 × 45 × 20

Other materials:
16 Scan bolts, nuts, and cover-heads
4 Table plates and $\frac{1}{2}$″ screws
$1\frac{1}{4}$″ × No. 6 Chipboard screws
1″ × No. 6 Countersunk woodscrews
Edging strip
Woodworking adhesive
Polyurethane

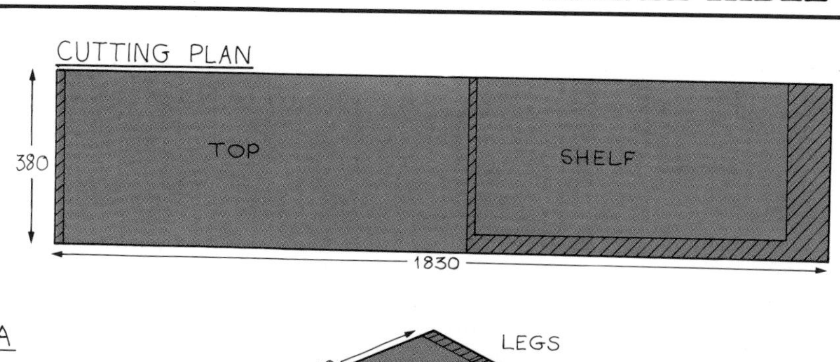

CUTTING PLAN
TOP
SHELF
380
1830

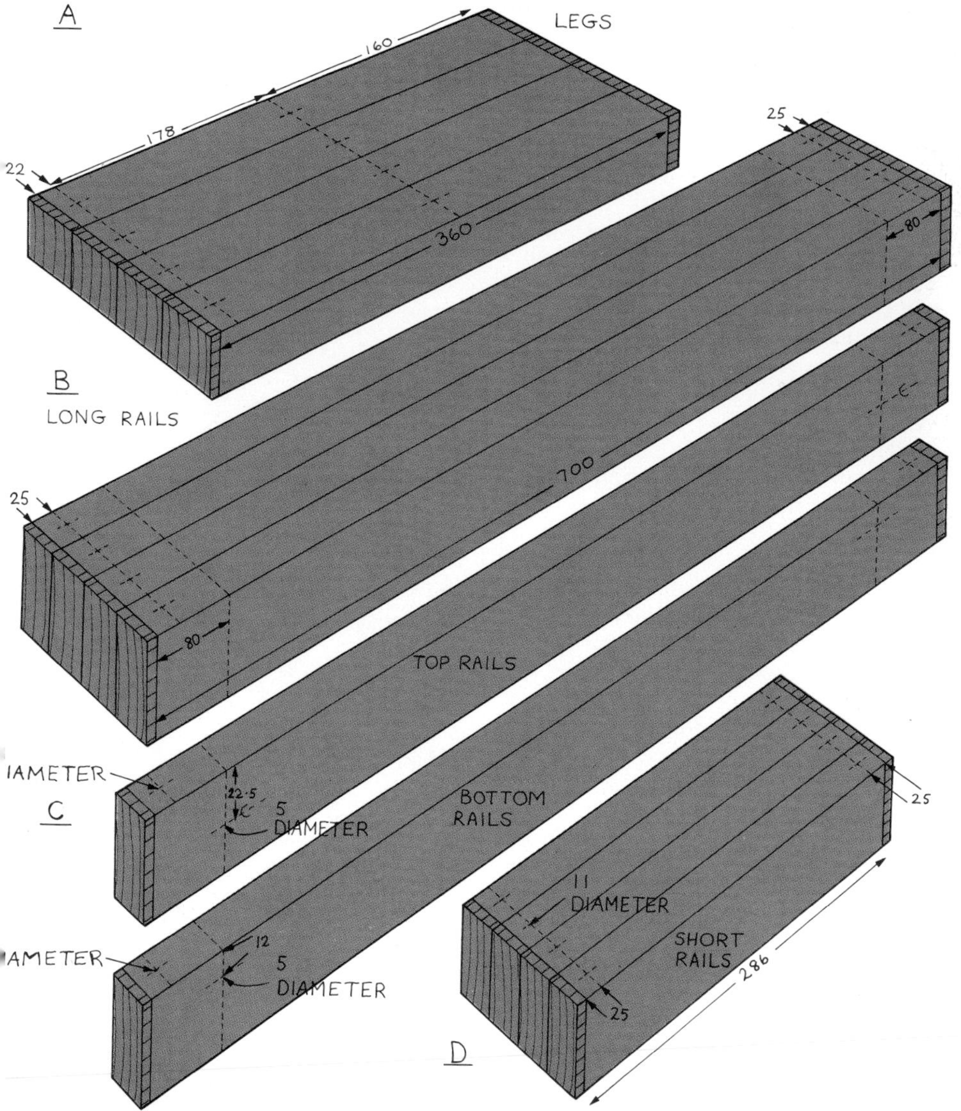

A
LEGS
160
178
22
360
25
80
B
LONG RAILS
25
700
80
TOP RAILS
IAMETER
C
22·5
5
DIAMETER
BOTTOM
RAILS
AMETER
12
5
DIAMETER
11
DIAMETER
SHORT
RAILS
286
25
25
D

top rail the hole for the Scan fitting is dead centre, but on the bottom rail this hole is 33mm from the top edge as in diagram C. Saw off the waste from each end. Bore and counter-bore. The four short rails are marked out as in diagram D.

The ends of all the long rails and all the short rails are prepared for the Scan fittings in exactly the same way. Bore the 11mm diameter holes to the depth required using a vertical drill-stand. The holes in the ends of all the rails are best bored by making a small work aid as shown in diagram E. All marking out should be done using a marking gauge. The work aid is made using the vertical drill-stand and care should be taken to ensure that this is made with a good degree of accuracy. The work aid is cramped on to the end of each rail and guides the drill into the exact position. Once each rail has been bored, clean it up using a plane and/or glasspaper and then round the corners as shown in diagram F.

Next, counter-bore all positions for the Scan nut as shown in diagram G. Now insert a 1″ No. 6 screw into the 3mm hole at each end of each long rail. Screw these in so that 6mm of shank is left standing. Then cut off the screw with a small hacksaw leaving two steel 'pegs' protruding. These pegs prevent the rails from twisting round the Scan bolt once the framework is assembled (diagram H). Tighten each joint into position. The steel pegs will then mark their positions on to the mating surfaces. Take the joint apart, and bore holes 6mm deep to accept the steel pegs. Reassemble using a woodworking adhesive. Finally, cut the Contiboard top and shelf to size. Using a 550mm handsaw with approximately 9 teeth to 25mm, cut the board on the waste side of your cut line. Chipboard can be planed, but it is best to work from both sides to the middle to prevent damage to the board edge. Veneer the edges with edging strip. Attach the top to the top rails with table plates and screw through the bottom cross-rails into the underside of the shelf. Finish with polyurethane varnish for a stain-resistant surface.

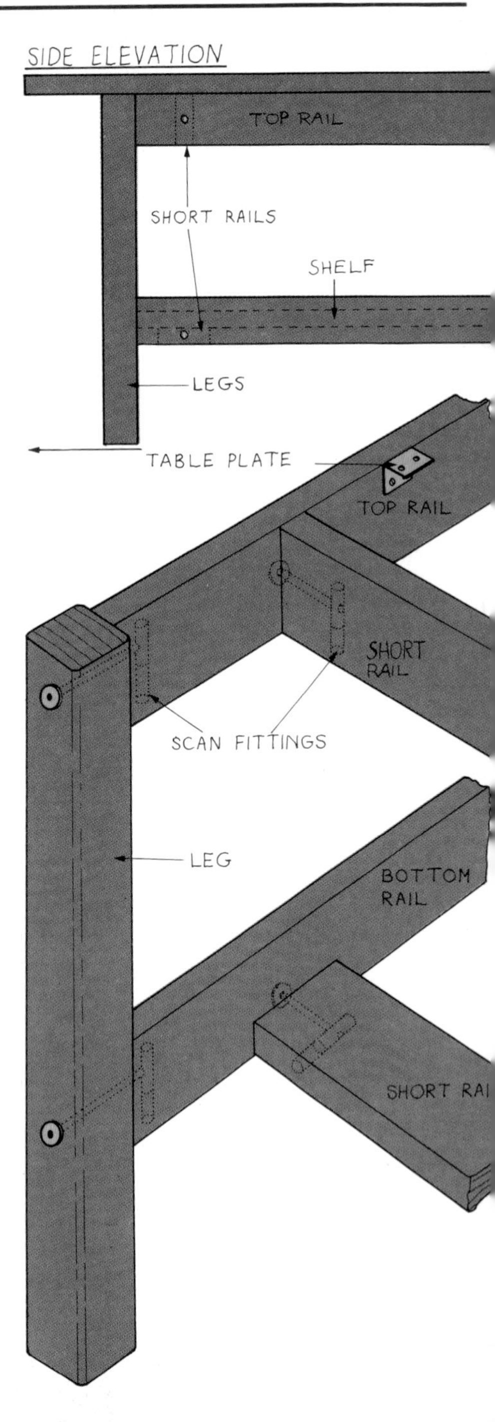

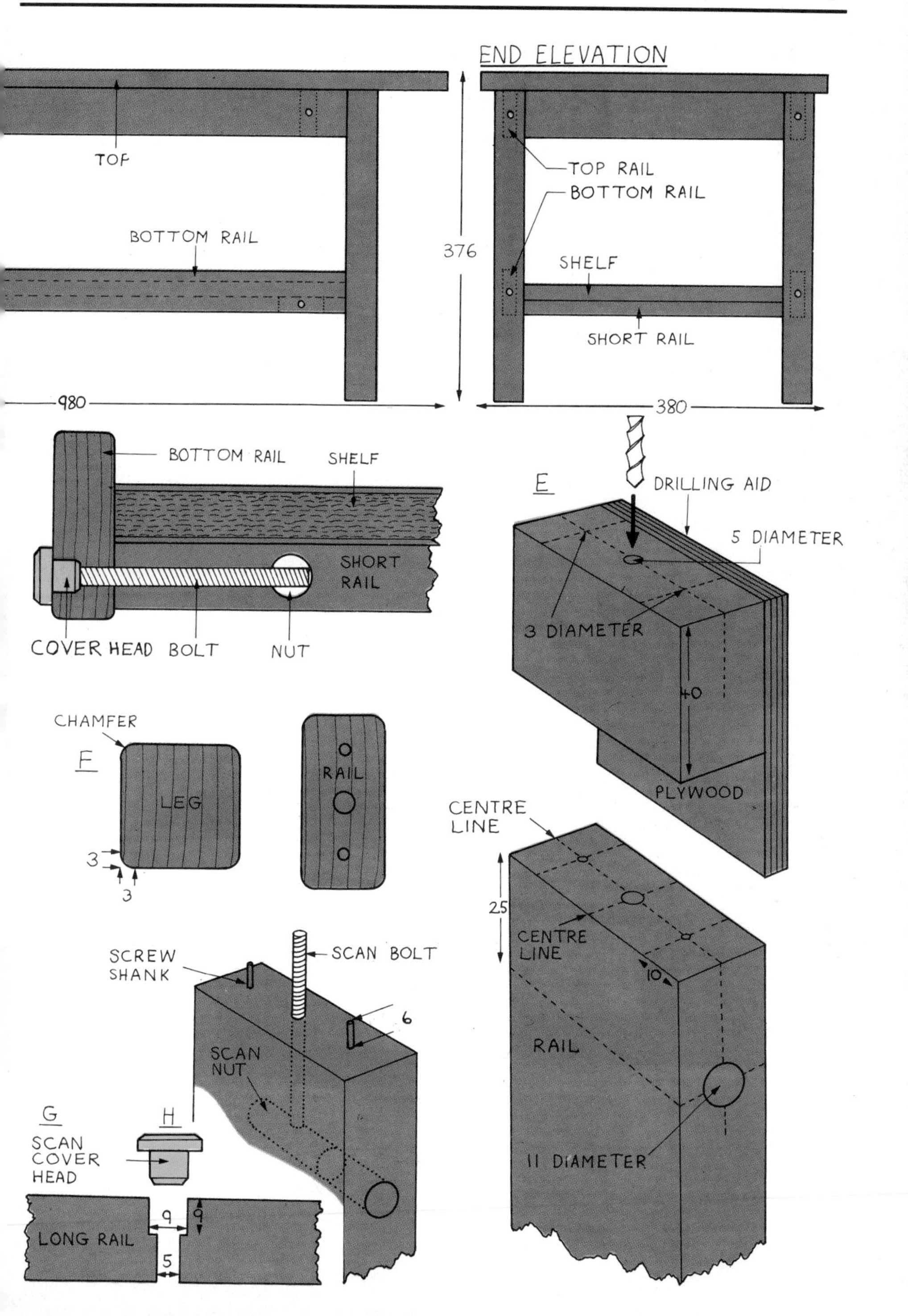
END ELEVATION
TOF
BOTTOM RAIL
376
980
TOP RAIL
BOTTOM RAIL
SHELF
SHORT RAIL
380
BOTTOM RAIL
SHELF
SHORT RAIL
COVER HEAD
BOLT
NUT
E
DRILLING AID
5 DIAMETER
3 DIAMETER
40
PLYWOOD
CHAMFER
F
LEG
3
3
RAIL
CENTRE LINE
25
CENTRE LINE
10
RAIL
11 DIAMETER
SCREW SHANK
SCAN BOLT
6
SCAN NUT
G
SCAN COVER HEAD
H
LONG RAIL
9
9
5

Made in pine-board with white, plastic-laminate faced boards for the interior of the cosmetic-storage section, this dressing-table makes an attractive piece of furniture. Of straightforward box construction, without any fancy joints or legs to shape and fit, it can be made with the minimum of tools. If you should choose veneered chipboard you will have to finish any exposed edges with iron-on edging strip.

Standard-width boards are used throughout, but in order to provide an overhang so that the lid can be lifted without having to have a handle fitted, the ends and centre panel have to be reduced slightly. The top of the unit is 1066mm long and the lid, which is 616 × 375mm, is cut from it.

The top, end and centre panels are all

Cutting list (millimetres)

Pine-board:	
Top	1066 × 450
Lid	615 × 375
Ends (2), Centre panel	647 × 447
Open shelf, Bottle-tray front	598 × 150
Top rail	1030 × 50
Back rail, Plinth	414 × 66
Upper drawer fronts (2)	412 × 150
Bottom drawer front	412 × 225
Contiplas:	
Shelf	598 × 182
Bottle tray	598 × 229
Hardboard:	
Back	1060 × 662
Drawer bases (3)	394 × 391

For timber drawers:
Plywood (9mm thick):

Bottom drawer back Bottom drawer sides (2)	400 × 203
Upper drawer backs (2) Upper drawer sides (4)	400 × 128

Timber:

Drawer base battens (6)	376 × 9 × 9
Drawer base battens (6)	391 × 9 × 9
Drawer runners (6)	400 × 12 × 12

Other materials:
4m Drawer profile, and corner connectors
1 Mirror, 450 × 300, and clips and screws
Adjustable braking stay or chain
Dowels 24 × 6 diameter
1 Piano hinge, 600 long, and screws
24 Joint blocks, and screws
$1\frac{1}{4}''$ × No. 6 Chipboard screws

Other materials (continued):
$\frac{5}{8}''$ × No. 6 Chipboard screws
$\frac{3}{4}''$ × Hardboard pins and Plastic screw-caps
Woodworking adhesive, Danish oil or other finish

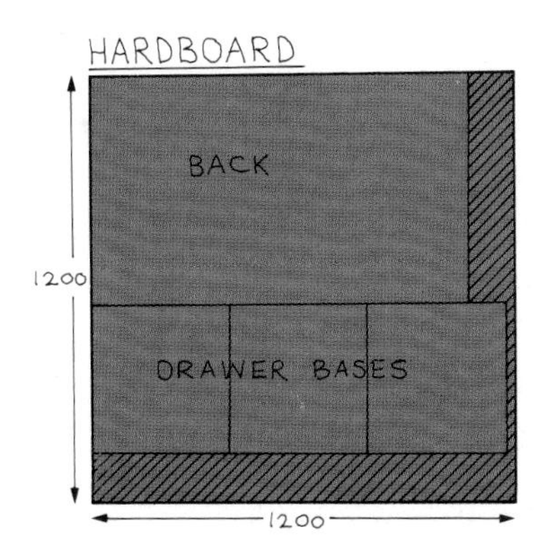

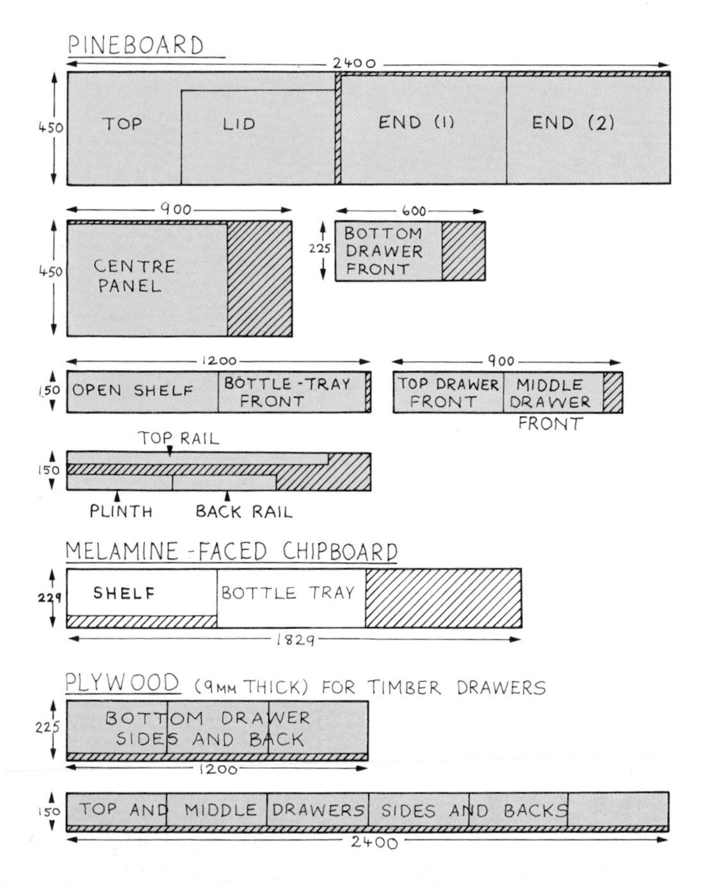

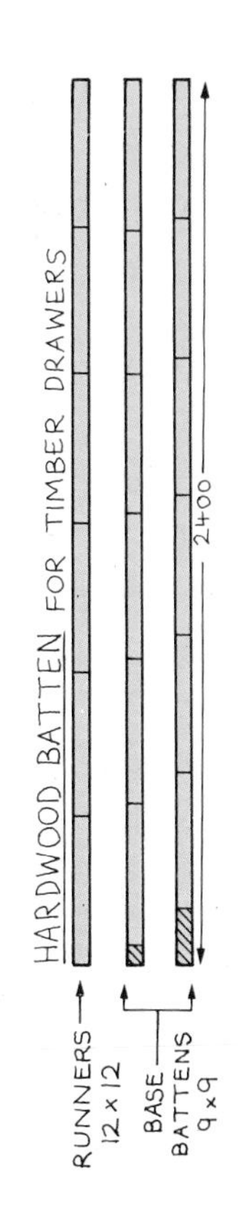

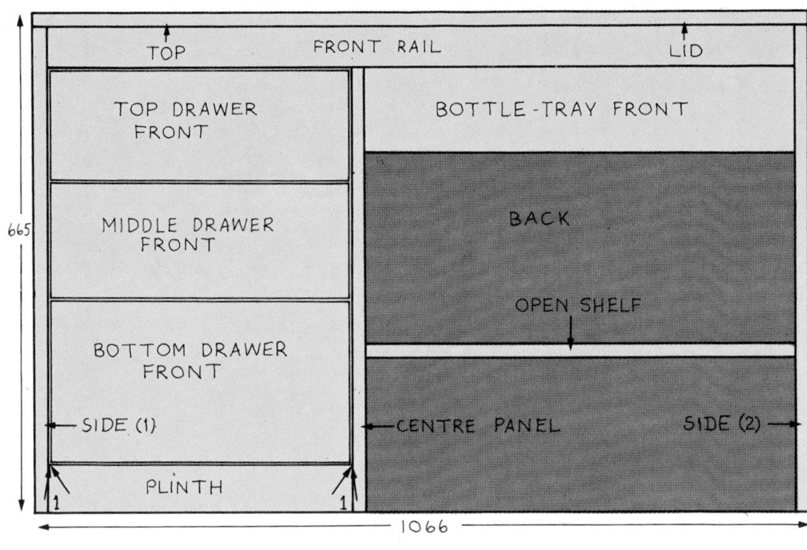

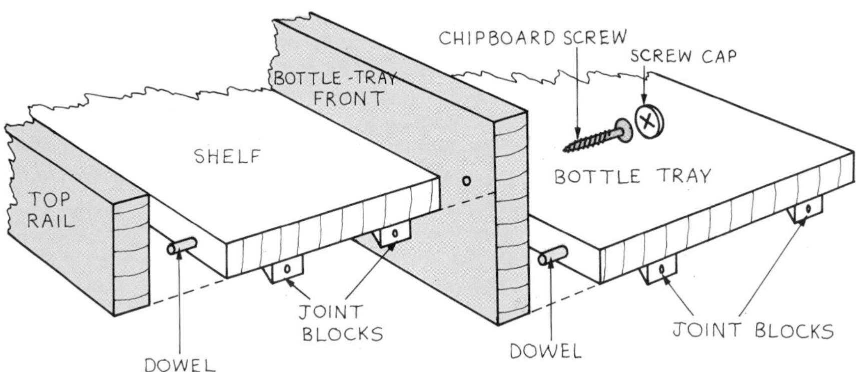

made from 450mm wide boards, the 647mm long end and centre panels being reduced in width by about 3mm so that the top will overhang them. The top of the centre panel has a notch 50mm wide and the same depth as the thickness of the board to take the 50mm wide top rail. Place the left-hand end and the centre panel together and mark on the edge the position of the top notch and the drawer fronts which are 150, 150 and 225mm wide. Allow 2mm clearance between the drawers and between the top drawer and the top rail, and 1mm at each side. Square these drawer marks accurately across the inside of the boards as they determine the positions of the drawer runners.

The drawers may be made either by using plastic drawer profile, following the manufacturer's instructions, or from timber as described below. In the latter case, the runners are made from 12mm square hardwood which is glued and screwed to the panels, with the bottom edge on the lines you have squared across them. The drawer runners also act as stops for the drawers so they must be positioned with the front end set back from the front edge of the board by the thickness of the drawer fronts.

Cut the 66mm wide plinth and the 50mm wide top rail from a 150mm wide board. The plinth is fixed by joint-blocks 18mm behind the edge of the end and centre panels, where it will butt up against the end

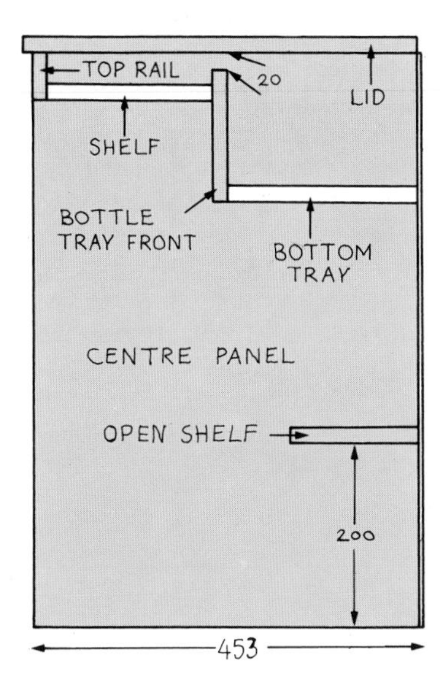

of the drawer runners. The top and the top rail are also fixed with plastic blocks, the top rail being flush with the panel edges.

A 150mm wide shelf is shown in the knee-hole of the dressing-table and this is positioned about 200mm up from the bottom of the panels. Its position is not critical and you can fit it where you think it will be most useful. Its main purpose is to hold the bottom of the end panel in place.

The shelf and bottle-tray inside the dressing-table are made from melamine-faced chipboard so that they can be cleaned easily and will not be damaged by spillage. As an extra precaution you should finish the sides of the inside of the dressing-table with hard gloss plastic-coating.

The shelf and bottle-tray can be assembled using screws with plastic caps, and dowels where the screws would show to the front. The whole assembly is fixed into place using plastic blocks underneath the shelf and tray where they will not be seen. A couple of short dowels into the front rail of the unit will provide extra support, and this joint should be glued as well. The top edge of this tray assembly must be about 20mm below the lid when it is closed, so that it will not foul the mirror when it is fixed in place.

At the back of the drawer section a rail of the same size as the plinth will be needed at the bottom to serve the same purpose as the small shelf under the dressing-table. This rail can be cut from the same board as the plinth and is fixed with two plastic blocks at each end.

At this stage the hardboard back can be fitted. It should be made about 6mm less than the width of the unit and 3mm less than the height so that when its edges are bevelled slightly, it will hardly be seen. Varnish or paint it and check that the unit is square by measuring the diagonals before fixing the back in place with glue and hardboard pins.

Plane smooth the cut edges of the lid and dressing-table top, and fix the piano hinge to the back edge of the lid so that it fits neatly back into the board from which it was cut and the grain pattern matches.

Support the lid by means of a chain, or a joint stay with a centre pivot. A standard mirror is fitted to the inside of the lid using the fixings provided with it.

The drawer fronts are made from the same type of board as the rest of the dressing-table, but the sides and backs are made from 9mm plywood. The drawer bottoms are made from hardboard or 3mm thick plywood.

The drawer fronts ought to be rebated to take the sides, but you can make simple butt-joints and fix the front to the sides using either joint-blocks or thin batten (9 × 9mm) glued and screwed to the sides and the front. Position the bottom of the sides 12mm up from the bottom of the front.

The back and the sides can be joined in the same way, but it is better to make a 3mm deep rebate in the sides.

Battens 9 × 9mm are also used to support the drawer bottoms. Glue and pin them round the inside of the drawer about 3mm above the bottom edge of the drawer sides so that they will not interfere with the sliding of the drawer on the runners. The bottom is then cut exactly square and rested on the battens where it is also glued and pinned. Fit handles as required.

DRAWING DESK

This pine-board desk features a top which can be tilted up to form a drawing-board and also a wide drawer which can take large sheets of drawing paper. When in use as a desk or table, the top is quite firm and flat. A secondary top or dust board of thin plywood or hardboard covers the drawer when the desk top is raised so that the contents will not get dirty.

Only one shelf is shown under the desk, for stiffening the structure, but more can be fitted if required. If the drawer is not needed for large sheets of paper, it can be partitioned to take small articles without them becoming mixed up with each other.

Two brass hinges are used to fix the top, but a piano hinge could be used to avoid having to cut recesses. Butt hinges are used for the support legs of the top, or flush hinges which do not need recessing.

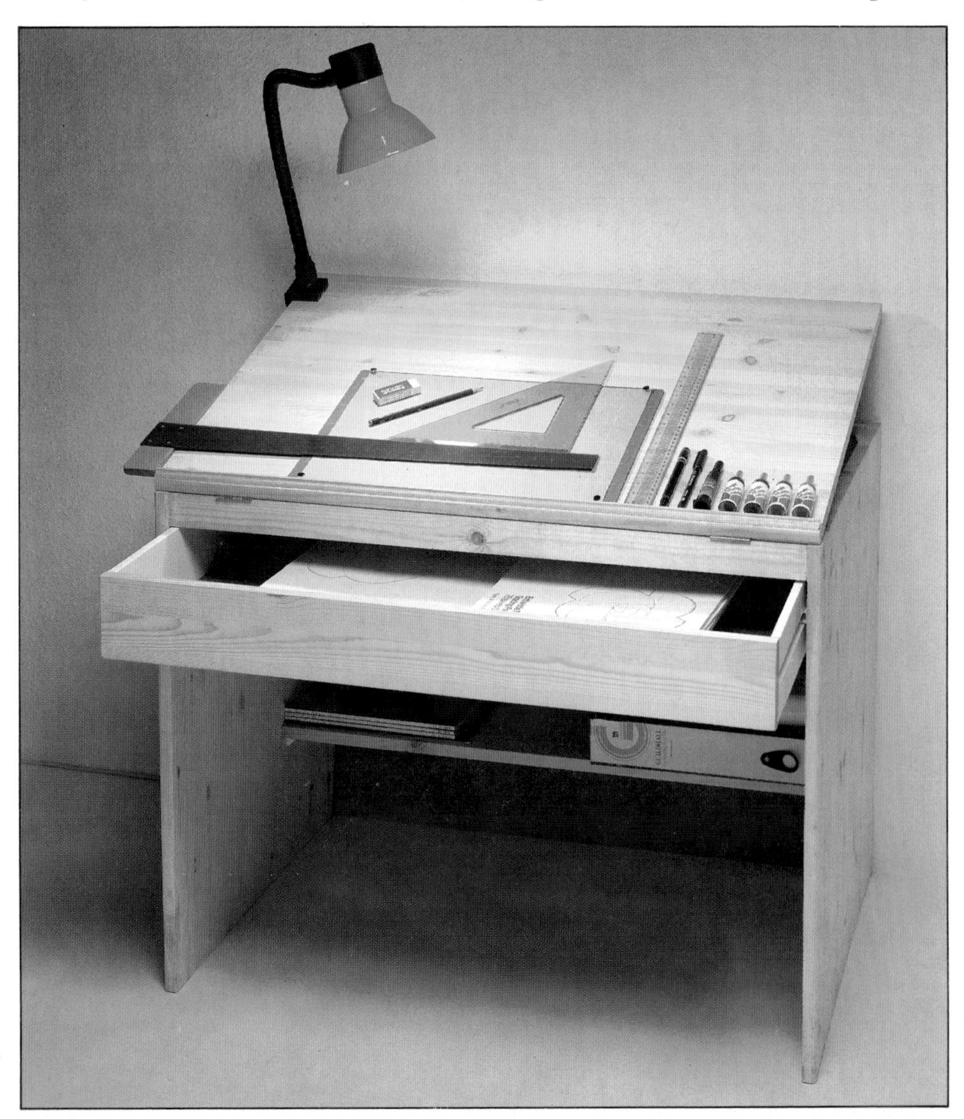

The side panels are 700 × 510mm and the top is 900 × 510mm. There is a narrow front-rail 40mm wide and a back rail 140mm wide; both of these and the shelf are 864mm long. The drawer front can also be cut off to this length, but the ends will require trimming to make it an easy fit when the cabinet has been constructed. When these parts have been cut square at the ends they can be assembled using joint-blocks. The small permanent type are most suitable for the wider back-rail and the shelf, where two will be required at each end. The front rail will have to be fixed with the two-part type

Cut a hardboard back to fit the unit so that it covers the back rail by about 25mm for nailing. If the side edges are bevelled slightly they will not show when the back is in place. Be careful to cut the hardboard exactly square as it will hold the desk unit square when it is glued and nailed into position. You can either paint the board before it is fixed or varnish it afterwards when the whole of the unit is given its finishing treatment.

Next, fix the battens for the dust board.

Cutting list (millimetres)

Pine-board:	
Top	900 × 510
Sides (2)	700 × 510
Shelf	864 × 150
Drawer front	862 × 100
Front rail	864 × 40
Back rail	864 × 140
Support legs	175 × 50
Hardboard:	
Back	896 × 575
Dust-cover	864 × 474
Drawer base	860 × 440
Timber:	
Dust-cover supports (2)	864 × 9 × 9
Dust-cover supports (2)	456 × 9 × 9

Other materials:
900mm Scotia moulding, 18 × 18
2 38mm Butt or flush hinges, and screws
2 50mm Butt hinges, and screws
2m 100mm Drawer profile and corner posts
1m Drawer base support moulding
1m Drawer runner profile
8 Joint blocks (rigid) and screws
2 Joint blocks (knock-down) and screws
Polyurethane or Danish oil

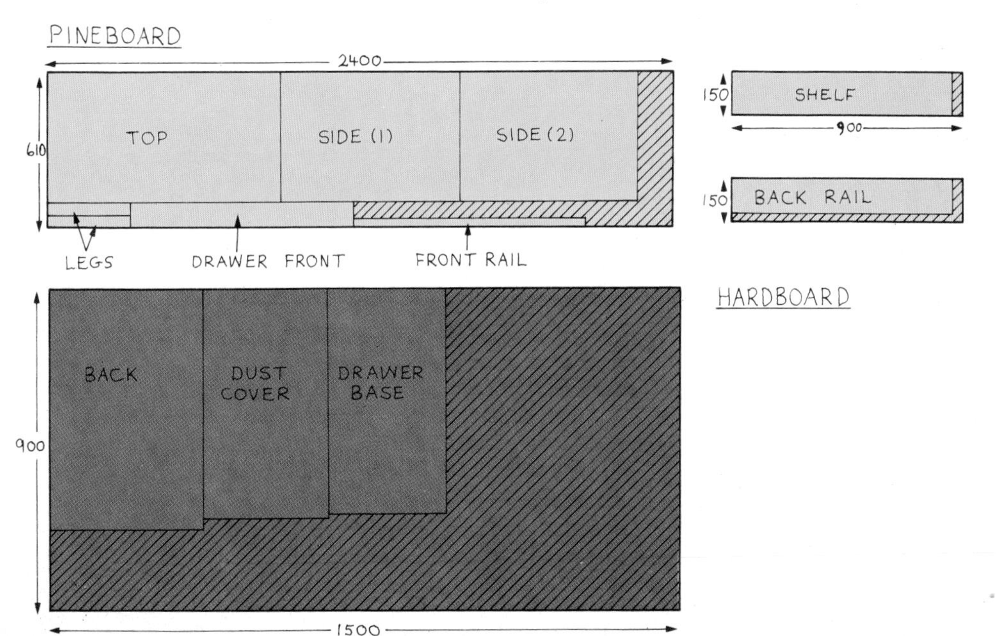

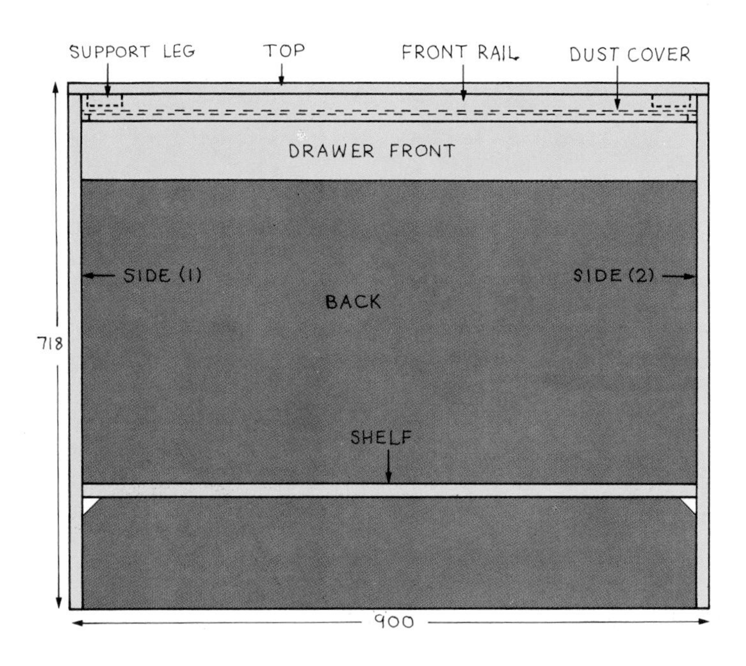
SUPPORT LEG
TOP
FRONT RAIL
DUST COVER
DRAWER FRONT
SIDE (1)
SIDE (2)
BACK
718
SHELF
900

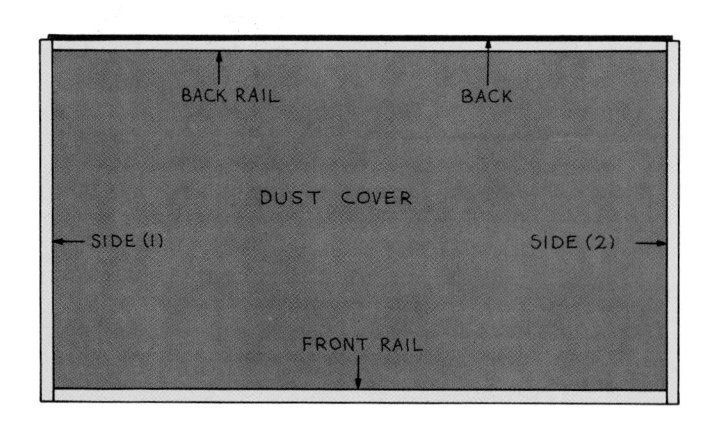
BACK RAIL
BACK
DUST COVER
SIDE (1)
SIDE (2)
FRONT RAIL

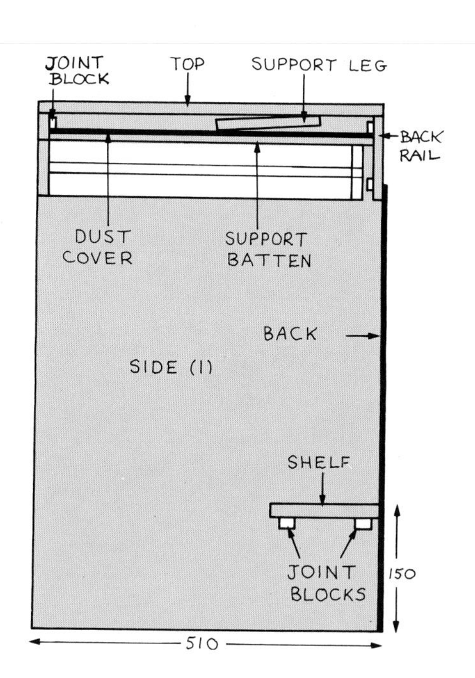

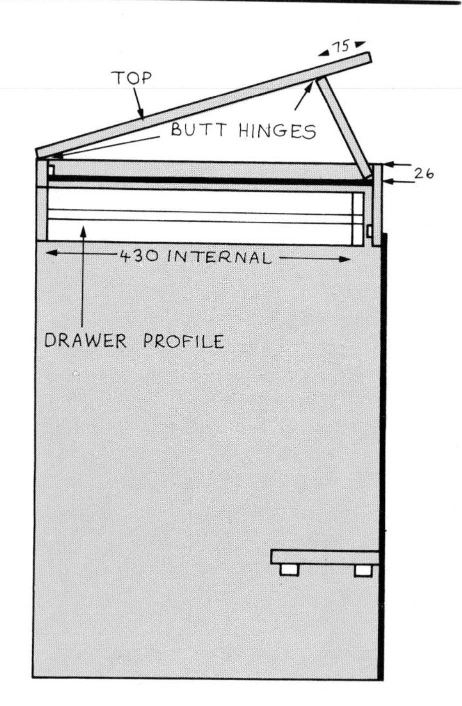

They are 9mm square and are pinned and glued with their undersides flush with the underside of the front rail. Then cut and fit the dust board, again remembering that it must be square because it will square up the whole unit when it is in place.

The supporting stays for the desk top are made from 50 × 18mm pineboard. Make them about 175mm long, and hinge them 75mm from the back edge of the top and 20mm in from the sides. They will then rest on the side supporting battens of the dust board and not just on the board which might bend and make the raised top unsteady.

These stays can be left in place, or taken off again to be refitted when the top has been hung on to the front edge of the desk. If butt hinges are used for the top, they should be let into the front rail with the knuckle projecting as little as possible. If a piano hinge is used, the front rail should be cut a little narrower to allow for the thickness of the closed hinge. A length of scotia moulding pinned and glued to the desk top, flush with the front edge, prevents pencils and other drawing equipment from sliding off when the top is in the raised position.

With the framework complete, attention can be turned to the drawer. This is 100mm deep and should have an internal dimension from front to back of 430mm so that it will take A2 size paper. It is constructed (following manufacturer's instructions) from plastic drawer profile and corner posts using the three-sided method. The drawer front is attached using wood-front corner fittings, after cutting to size and fitting the hardboard base. The front edge of the base is supported on a drawer base support, screwed to the inside of the drawer front; this avoids having to cut a groove in the board. Mount the runner mouldings to give a clearance of 2mm between the drawer front and the top rail, and set them in from the front edge by 18mm plus the thickness of the front corner fittings. The desk, in its complete form can now be finished off with a protective paint or oil.

This versatile desk unit is made from pine-board using joint-blocks for most of the joints so that it can be easily constructed using only a basic tool-kit. Counter-flap hinges are used so that the work-top will fold back and the front portion will lie flat on the back portion when in the closed position. Butt hinges could be used as an alternative.

A foot rest or rail is fitted between the front legs to make them firm. The plastic blocks used to fix it are screwed to the front of the rail when it is in the down position so that they will be out of sight when raised.

No catches are needed to hold the legs when in the raised position as their own weight will be sufficient. However, if you would prefer them fixed, perhaps so that young children cannot pull them over, you can fit a bolt under the wall shelf-unit to shoot into the desk leg.

The wall shelf-unit is entirely separate from the folding base-unit and although it is shown with a single shelf, its storage space can be designed to suit your own requirements. Only the outer dimensions of the unit cannot be varied because the bottom shelf must be situated high enough up the wall to clear the work-top when it is folded, and the top shelf of the unit must fit underneath the foot rest when the legs are raised. If preferred, it can be fitted with doors, either solid wood or glass. This applies to the lower unit too, which is also shown as a simple shelf-unit. These shelves can be made adjustable and doors can be fitted. They would have to be sliding doors as hinged doors would not open because of the position of the foot rest when the desk is open. Both the top and the bottom unit are secured to the wall when complete.

Start by cutting the pine-board tops to size, followed by the two sides and the two legs. The length of these is 720mm less the thickness of the tops. Then set out one pair of sides for the shelves, if they are to be fixed, or if bookcase strips for adjustable shelves are to be used screw them into place, allowing for the top, plinth and bottom shelf.

At this stage the base unit could be assembled, but it would be better if the two halves of the top were hinged together first. This is done most easily on a bench or work table, even if the two halves are separated again afterwards for ease of assembly of the units.

Counter hinges are made of solid brass and have a double throw action which allows the flap to lie flat on the top of the fixed section. No special skill is needed to fit them – patience in marking out the recesses and chiselling them out is more important. Three hinges are used, one in the centre of the flap and the other two about 150mm from each end. Put each hinge in place and mark the position of the knuckle. Cut this recess so that the recess for the leaf can be marked using a sharp pencil or, better still, the pointed blade of a setting-out knife. Clamp the flap to your work-bench and, using a sharp chisel, cut out a little at a time so that the hinge leaf fits tightly into the recess. When properly hung the two flaps should have only the merest hint of a gap between their edges when they are opened up. Binding will occur when the flaps are folded if the hinges are let too far into the wood; one flap should lie on top of the other quite easily.

Having hinged the flaps successfully, the base unit can be constructed. It is simply joined together with joint-blocks using two to each joint at the top and bottom, and also for the shelves if they are to be fixed. The plinth is fixed in the same way, with a plastic block at each end and two evenly spaced under the bottom shelf.

The unit is held square by the hardboard back which must be cut accurately. It can be made about 4mm shorter than the width of the unit and the edges can be bevelled so that it does not show at the sides. The top edge is left square and fits flush with the top of the unit. The back will hold the unit 3mm or so clear of the wall so that when the legs are folded up they too will be just clear of the wall.

Cut a solid-timber foot-rest to the same length as the top, less twice the thickness of the legs. The leg unit can be assembled in the same way as the base by using joint-blocks, but as the underside of the flap becomes the top when the legs are folded up, a triangular wooden-fillet, glued and screwed to the legs and the flap, would look better. An alternative would be to use a piece of brass or aluminium angle-strip which can also be used instead of joint blocks to fix the foot-rest. When the two units have been completed the hinges are refitted.

As the wall unit has to fit neatly between the legs and has to clear both the arc of the flap and the foot-rest, it is best to check all measurements before starting to cut the boards to length.

A simple shelf-unit is shown here, but it can be divided up in whichever way will be most useful. Lay-on doors using concealed hinges can be fitted, either when the unit is made or later if they become necessary. Lay-on doors would not be suitable for the base unit as they would stand proud of the top when the legs were folded up.

Cut the top and bottom to the full length, but cut the sides to the thickness of the top and bottom less than the full length. Use two plastic blocks at each of the joints and under the shelf, which is cut twice the thickness of the boards less than the length of the top.

Cutting list (millimetres)

Pine-board.	
Top Work-top	1000 × 225
Sides (2)	702 × 225
Desk legs (2)	702 × 150
Bottom	964 × 225
Adjustable shelves (2)	956 × 225
Wall-unit top Wall-unit bottom	900 × 150
Wall-unit sides (2)	300 × 150
Wall-unit shelf	864 × 150
Hardboard:	
Base-unit back	996 × 650
Wall-unit back	896 × 332
Timber:	
Foot-rest	964 × 75 × 50
Plinth	964 × 50 × 22

Other materials:
3 Counter hinges or butt hinges
4 Bookcase strips, 600 long, and 8 supports
28 Joint blocks, and screws
450mm Triangular fillet or metal angle
$\frac{3}{4}$" Hardboard pins
Woodworking adhesive
Polyurethane or Danish oil

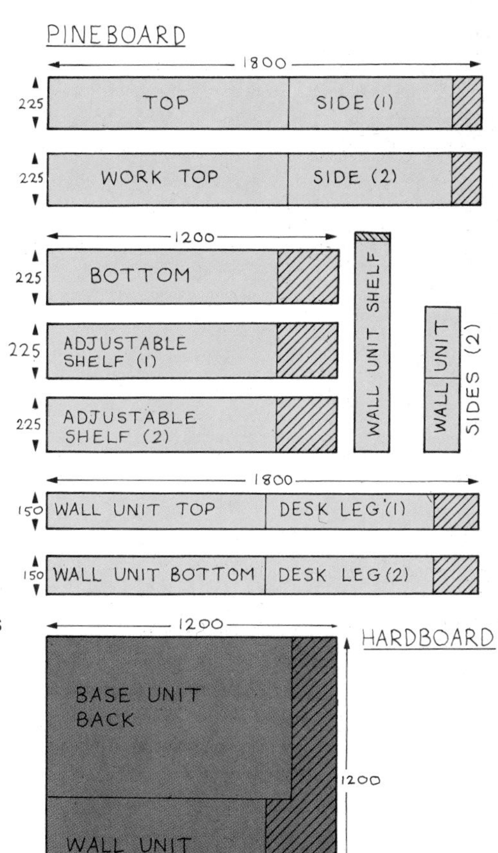

NOTES
GAMES COMPUTING

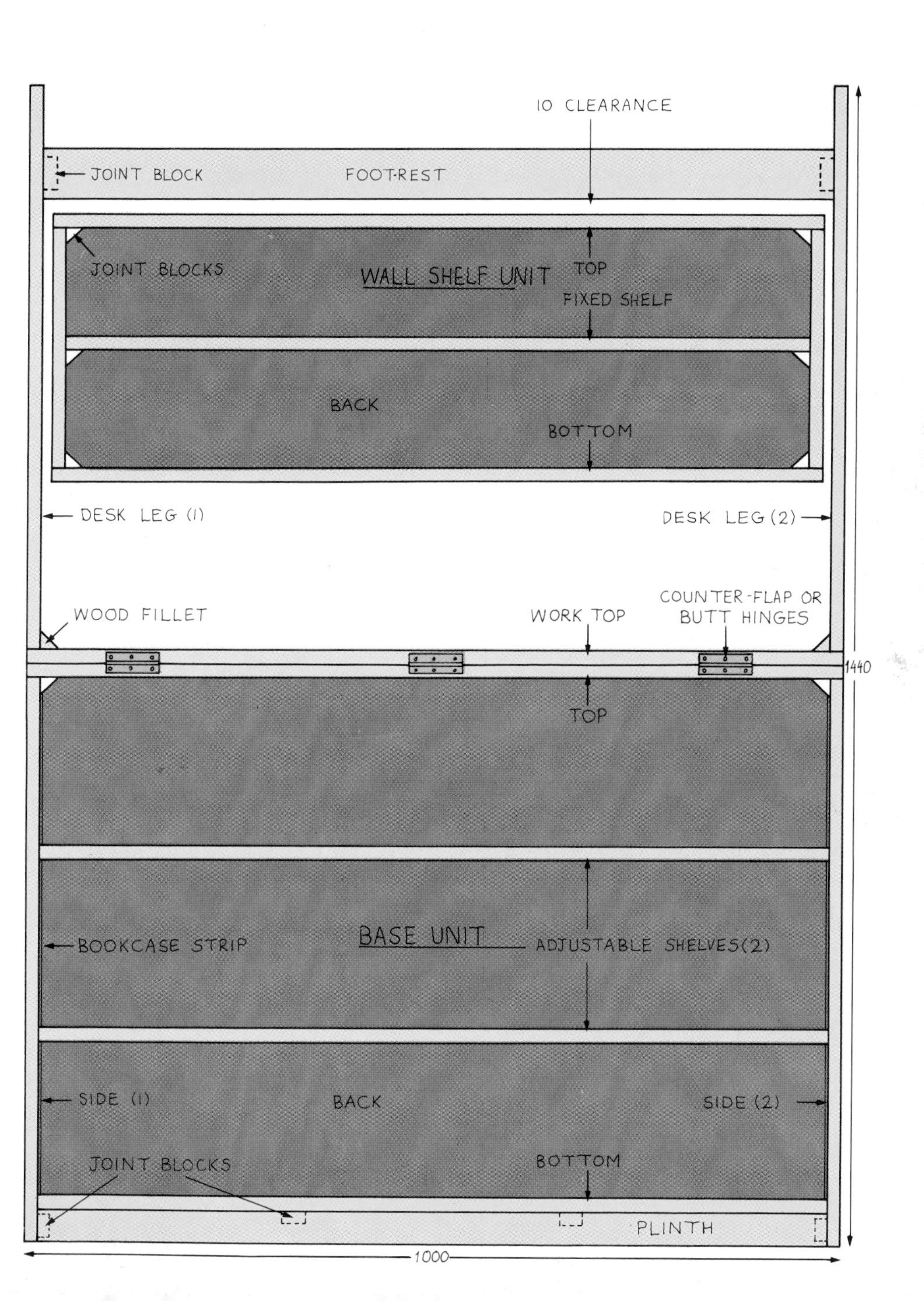
10 CLEARANCE
JOINT BLOCK
FOOT-REST
JOINT BLOCKS
WALL SHELF UNIT
TOP
FIXED SHELF
BACK
BOTTOM
DESK LEG (1)
DESK LEG (2)
WOOD FILLET
WORK TOP
COUNTER-FLAP OR
BUTT HINGES
1440
TOP
BASE UNIT
BOOKCASE STRIP
ADJUSTABLE SHELVES (2)
SIDE (1)
BACK
SIDE (2)
JOINT BLOCKS
BOTTOM
PLINTH
1000

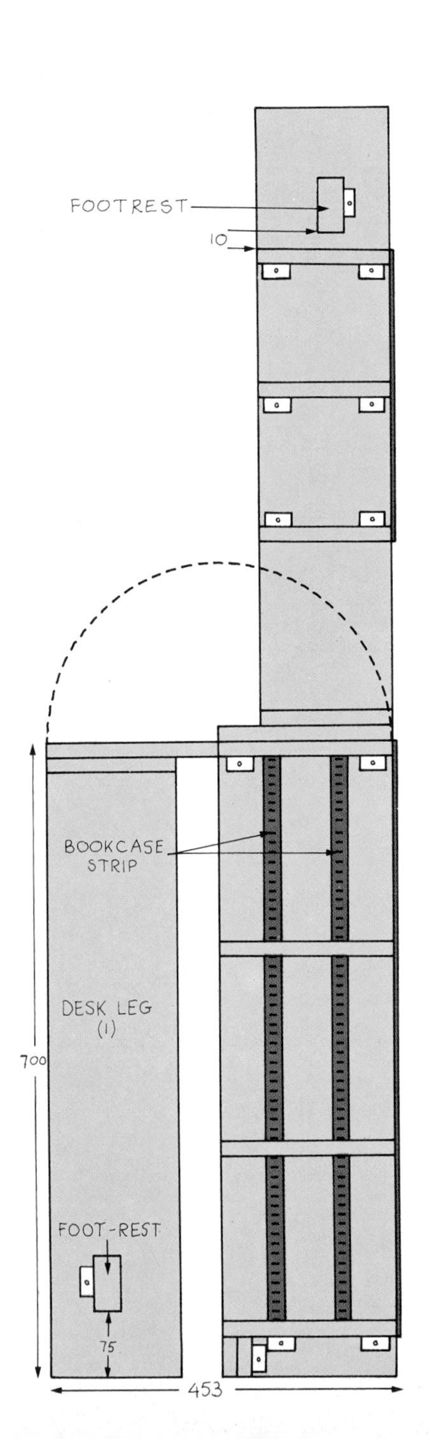

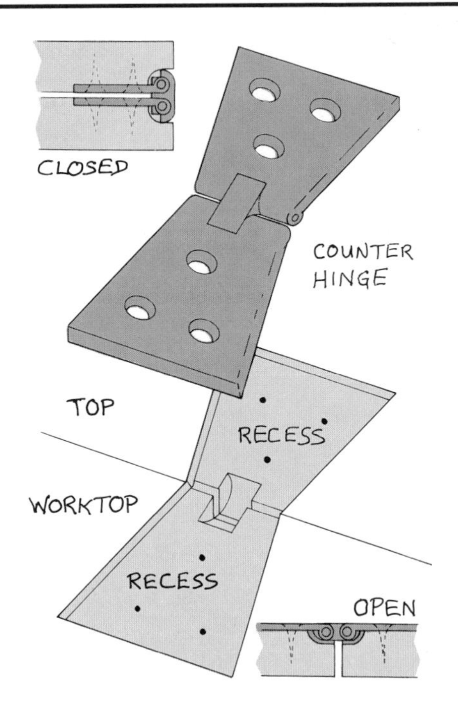

The back for this unit can be in hardboard or thin plywood. Cut it 4mm less than full size and ensure that it is exactly square as it will be used to hold the shelf unit square. If the unit is not square it may foul the legs when they are raised.

The type of fixing to the wall will depend on the weight which the unit is likely to have to carry. Paperbacks are not very heavy, but hardbacks can be a considerable weight. For light weights you can make a screw-fixing through the hardboard back of the unit. A plywood back will support heavier materials. Another method of supporting heavier materials is to fit a couple of mirror plates into the edge of the unit top. Recess them flush with the edge of the board, position the plates so that they point downwards, then when the hardboard back is in place, screw through the mirror plates and the backboard into the wall. Ensure that the screws get a good 38mm grip in solid brickwork and use a plastic or fibre plug for the best results. For extra support, screw a 50 × 25mm batten to the wall beneath the unit.

Acknowledgments
The publishers wish to acknowledge the help of the Timber Research and Development Association for the use of photographs (p.7), Lindsey Edwards and Ron Kidd for constructing the projects, and Ronnie Rustin for demonstrating French polishing.

The following organisations kindly supplied equipment and materials:
Aaronson Brothers; Black and Decker; DIY Timber; General Woodwork Supplies; Rustins; Stanley Tools.

Special photography: Chris Linton; Simon de Courcy Wheeler.

Illustrations: Will Giles; Janos Marrfy.